Anonymous

The Anatomical Remembrancer, or, Complete Pocket Anatomist

Containing a concise description of the structure of the human body

Anonymous

The Anatomical Remembrancer, or, Complete Pocket Anatomist
Containing a concise description of the structure of the human body

ISBN/EAN: 9783337370596

Printed in Europe, USA, Canada, Australia, Japan

Cover: Foto ©berggeist007 / pixelio.de

More available books at **www.hansebooks.com**

THE ANATOMICAL REMEMBRANCER,

OR,

Complete Pocket Anatomist:

CONTAINING

A CONCISE DESCRIPTION

OF

THE STRUCTURE OF THE HUMAN BODY.

Third Edition.

WITH

CORRECTIONS AND ADDITIONS

By C. E. ISAACS, M.D.,

Demonstrator of Anatomy in the University of New York.

NEW YORK:
WILLIAM WOOD & CO., 27 GREAT JONES STREET.
1871.

Entered according to Act of Congress, in the year 1871,
BY WILLIAM WOOD & COMPANY,
In the Office of the Librarian of Congress, at Washington, D.C.

THE NEW YORK PRINTING COMPANY,
205, 207, 209, 211 *and* 213 *East Twelfth Street.*

PREFACE TO THE FIRST EDITION.

THE sole object of this little Manual is to recall to the mind of the Student in Anatomy the information he may have acquired, either by actual dissection, or by the perusal of works which profess to treat more fully on the subject. The Author, therefore, presumes it will be found highly useful to those who are preparing for examination at the Royal Colleges of Surgeons and Societies of Apothecaries.

PREFACE TO THE FOURTH EDITION.

IN the attempt to introduce the more important modern discoveries in Anatomy, the Editor has experienced difficulty in keeping this work to its proper limits: he hopes, however, that it will be found to contain sufficient to recall to the Student in Anatomy that information which he has acquired in the Dissecting Rooms. He has endeavored to simplify the description of certain regions, which conflicting inaccuracies have rendered almost unintelligible. Foremost amongst these stands "the perineum," which is justly regarded by the Student as an incomprehensible riddle; next comes the anatomy of the brain; thirdly, the muscles of the back. The minute structure of bone and of some of the more important glandular structures has for the first time been introduced, such information being required by the examining Boards of the Profession.

October, 1850.

PREFACE TO THE SECOND AMERICAN EDITION.

THIS work, as its name indicates, is intended as a Pocket Companion, and aid to the memory of the Student, and especially while engaged in the details of Practical Anatomy.

Several errors in the last edition have been corrected, and some additions made to the present, which seemed to be useful, and demanded by the advanced state of the science. Hernia is one of the subjects which usually present the greatest difficulty to Medical Students : such a description has therefore been given of its anatomy and dissection, as from long experience I have found to be most easily comprehended, avoiding all unnecessary complexity and detail, or reference to discordant statements of different authors.

It is hoped that this book, in its present form, will be of increased utility to those for whose use it was especially designed.

C. E. ISAACS, M.D.

INDEX.

OSTEOLOGY.

	PAGE
MINUTE STRUCTURE OF BONE	1
THE SPINAL COLUMN	2
Common characters of a true vertebra	2
Characters of the lumbar vertebræ	3
Deviations	4
Characters of the dorsal vertebræ	4
Deviations	5
Characters of the cervical vertebræ	5
Deviations	6
False vertebræ	9
Os sacrum	9
Os coccyx	10
THE THORAX	11
Common characters of a rib	11
Deviations	12
The sternum	12
THE PELVIS	13
Os ilium	14
Os pubis	15
Os ischium	16
THE BONES OF THE SUPERIOR EXTREMITY	17
THE BONES OF THE LOWER EXTREMITY	28
THE SESAMOID BONES	40
THE SKULL AND FACE	40
Frontal bone	41
Parietal bone	43
Occipital bone	44

	PAGE
Temporal bone	45
Ethmoid bone	48
Sphenoid bone	50
Malar bone	52
Superior maxillary bone	53
Palate bone	55
Inferior spongy bone	57
Lachrymal bone	57
Nasal bones	58
Vomer	58
Inferior maxillary bone	59
Os hyoides	60
THE CRANIAL AND FACIAL SUTURES	60
THE ORBITS	62
THE TEMPORAL FOSSÆ	63
THE ZYGOMATIC FOSSÆ	63
THE PTERYGO-MAXILLARY FISSURES	63
THE ARTICULATIONS	64
Temporo-maxillary articulation	64
Occipito-atlantoid articulation	65
Occipito-axoid articulation	65
Atlanto-axoid articulation	66
Common vertebral articulation	66
Costo-spinal articulation	67
Chondro-sternal articulation	68
Lumbo-sacral articulation	68
Ilio-sacral articulation	69
Sacro-coccygeal articulation	70
Pubic articulation	70
Sterno-clavicular articulation	70
Scapulo-clavicular articulation	71
Coraco-clavicular articulation	71
Ligaments of the scapula	71
Humero-scapular articulation	72

	PAGE
Humero-cubital articulation	72
Superior radio-ulnar articulation	73
Inferior radio-ulnar articulation	74
Radio-carpal articulation	74
Carpal articulations	75
Carpo-metacarpal articulations	76
Metacarpo-phalangeal articulations	76
Inter-phalangeal articulations	76
Ilio-femoral articulation	76
Femoro-tibial articulation	77
Tibio-fibular articulation	79
Articulation of the ankle	79
Articulations of the tarsus	80
Tarso-metatarsal articulations	81
THE MUSCLES	81
Muscles of the head	81
Muscles of the external ear	82
Muscles of face	82
Muscles of lower jaw	84
Muscles on anterior and lateral parts of the neck	84
Muscles of the pharynx	86
Muscles of the palate	87
Muscles of the larynx	87
Deep muscles on anterior and lateral parts of the neck	88
Muscles of the thorax	89
Muscles of the back	90
Muscles of shoulder and arm	94
Muscles of fore-arm and hand	96
Muscles of abdomen	100
Deep muscles of abdomen	102
Muscles of male perinæum	103
Muscles of female perinæum	104

	PAGE
Muscles of inferior extremity	105
Muscles of hip	107
Muscles on back of thigh	108
Muscles on anterior and external part of leg	109
Muscles on outer part of leg	110
Muscles on back of leg	110
Muscles of foot	111
Muscles of orbit	113
Muscles of internal ear.	114
THE BRAIN AND ITS MEMBRANES	115
Dura mater	115
Sinuses	116
Tunica arachnoidea	118
Pia mater	119
Cerebrum	119
Cerebellum	127
Medulla oblongata	127
Base of the brain	128
Origins of the cerebral nerves	130
Distribution of the cerebral nerves	131
Ganglions in connexion with the fifth pair	134
Spinal nerves	139
Cervical plexus	140
Brachial plexus	141
Dorsal nerves	144
Nerves of Wrisberg	145
Lumbar plexus	145
Sacral plexus	147
Sympathetic nerves	148
Cervical ganglions	149
Cardiac nerves	150
Cardiac plexus	150

	PAGE
Thoracic ganglions	151
Semi-lunar ganglions	151
Solar plexus	152
Renal plexus	152
Inferior mesenteric plexus	152
Lumbar ganglions	152
Sacral ganglions	152
Ganglion impar	153
THE THORAX AND ITS CONTENTS	153
Pleuræ	154
Anterior mediastinum	155
Middle mediastinum	155
Posterior mediastinum	155
THE LUNGS	156
Trachea and its Ramifications	157
THE HEART AND PERICARDIUM	159
THE ARTERIES	165
THE VEINS	177
THE DIGESTIVE APPARATUS	183
Mouth	183
Teeth	183
Pharynx	184
Œsophagus	184
Stomach	185
Duodenum, jejunum, and ilium	187
Cœcum	189
Colon	189
Rectum	189
Parotid glands	190
Submaxillary glands	191
Sublingual gland	191
Liver	191
Pancreas	195
Spleen	195

	PAGE
THE URINARY APPARATUS	196
Kidneys	196
Ureters	198
Urinary bladder	199
THE PERITONEUM	201
THE MALE ORGANS OF GENERATION	205
Testicles	205
Spermatic cord	208
Vesiculæ seminales and prostate gland.	209
Cowper's glands	210
Penis	210
Urethra	211
THE FEMALE ORGANS OF GENERATION	213
Mammæ	216
THE ORGANS OF THE SENSES	217
Organ of touch	217
Organ of smell	218
Organ of taste	220
Organ of vision	220
Appendages of the eye	224
Organ of hearing	226
Absorbent system	231
PECULIARITIES OF THE FŒTUS	233
Cervical fascia	234
Superficial fascia of the abdomen	235
Fascia transversalis and fascia iliaca	236
Deep perineal fascia, or triangular ligament of the urethra	237
Fascia of upper extremity	239
Fascia lata	239
THE LARYNX	242
THE THYROID BODY	245
HERNIA	246

THE ANATOMICAL REMEMBRANCER.

MINUTE STRUCTURE OF BONE.

Bone is composed of gelatine and phosphate, and carbonate of lime. It is surrounded by a fibrous vascular membrane (periosteum): the long bones are lined by an internal membrane called medullary. The minute vessels of bone, all of small size, run in canals called Haversian; from the Haversian canals radiate minute tubes, which abstract the phosphate of lime from the blood; they are called "calcigerous" tubes, or canaliculi, and terminate in bone cells, or lacunæ. If bone be cut transversely, it seems under the microscope composed of rods, placed side by side, in the centre of each of which is a Haversian canal, sending off its radiating calcigerous tubes, and surrounded by concentric rings of bone cells.

OSTEOLOGY.

The assemblage of bones, constituting the framework of the human body, is termed the skeleton; it is divided into head, trunk, superior and inferior extremities. The whole number of bones is 211.

The trunk consists of the spine or vertebral column, and includes the thorax and the pelvis.

The Spinal or Vertebral Column, situated in the posterior part of the trunk, supports the head, and is supported by the pelvis. The bones which enter into its formation are called vertebræ, of which there are two classes, the *true* and the *false.*

The true vertebræ, twenty-four in number, are subdivided into three classes, viz., seven *cervical,* twelve *dorsal,* and five *lumbar.*

The false vertebræ are coalesced to form two pieces, viz., the os sacrum and os coccyx, both of which bones enter into the formation of the pelvis. The sacrum is composed of five and the coccyx of four pieces of bone.

COMMON CHARACTERS OF A TRUE VERTEBRA.

A ring of bone, the opening of which is called the spinal or vertebral foramen.

The body, a mass of bone placed anterior to the ring, thick, spongy, and presenting many small holes for blood-vessels.

Laminæ, two lateral plates which pass backwards from the posterior part of the body, forming the sides of the ring, and terminating posteriorly in the *spinous process,* from the existence of which the vertebral column has been called *spine.*

Two transverse processes, which pass outwards from the sides of the laminæ.

Four articular processes, two upon the upper and two upon the lower surface of each vertebra, situated at the roots of the transverse processes, and articulating with the vertebra above and below.

All these processes differ from the body in being formed of more compact bony texture.

Four notches, two above and two below, which are formed by the laminæ being grooved out where they join the body. Each of these, with the corresponding notch above and below, forms a lateral hole, the intervertebral foramen, for the exit of the spinal nerves and the entrance of blood-vessels.

CHARACTERS OF THE LUMBAR VERTEBRÆ.

1*st*. They are the *largest* of the three classes.

2*nd*. *The bodies* are very broad transversely, of oval form, deepened upon their upper and lower surfaces by a more compact lamina of bone, which, projecting beyond their bodies, renders them slightly concave from above downwards upon their fore part.

3*rd*. *The spinous processes* are broad, thick, and short.

4*th*. *The transverse processes* are long, thin, and horizontal, and are regarded by some anatomists as abdominal ribs.

5*th*. *The superior articulating surfaces* are oval, concave, and look inwards and backwards; *the inferior* being oval and convex, directed outwards and forwards; from each of the superior articular processes, which are wider apart than the inferior, there projects backwards the "tubercle."

6*th*. *The vertebral foramen* is of triangular shape, and larger than in the dorsal vertebræ.

7*th*. *The notches*, particularly the inferior, are very large, and form larger foramina than at any other part of the spine.

DEVIATIONS.

The last lumbar vertebra has its body cut off obliquely upon its sacral aspect, so that it is much thicker before than behind. Its transverse processes also are short and rounded.

CHARACTERS OF THE DORSAL VERTEBRÆ.

1*st*. They are intermediate in size between the cervical and lumbar vertebræ.

2*nd*. *Their bodies* are thicker behind than before, and more convex transversely, assuming upon their surfaces a triangular rather than an oval form. On either side, at the upper and inferior margins each body presents two small depressions, the upper being the larger, which with the intervertebral cartilage and the contiguous vertebra, form depressions for lodging the heads of the ribs.

3*rd*. *The spinous processes* are long, prismatic, tuberculated at their extremities, and directed obliquely downwards.

4th. The transverse processes are long, and directed backwards; on the extremity and anterior aspect of each is an oval articular surface for the tubercle of the rib.

5th. The articular processes are nearly vertical, the superior looking backwards, the inferior forwards.

6th. The vertebral foramen is smaller than in the cervical or lumbar vertebræ, and is of oval shape.

DEVIATIONS.

The *first dorsal vertebra* has a full depression for the head of the first rib, besides the half depression for the second; its body is longer in the transverse direction, its spinous process is strong and horizontal, and its articular processes are oblique.

The 11*th* and 12*th* have each a full depression upon the body for the corresponding rib, but want the depression on the transverse processes; the 11*th* and 12*th* have very short transverse processes, and the 12*th* resembles a lumbar vertebra in the shape of its body, and of its inferior articular processes.

Characters of the Cervical Vertebræ.

1*st*. They are the *smallest* of the three classes.

2*nd. Their bodies*, deeper before than behind, and long transversely, are concave from side to side upon their upper surface, and concave from before backwards upon their lower.

3rd. The laminæ are broad and thin.

4th. The spinous processes are short, horizontal, and bifid.

5th. The transverse processes are bifid and short, grooved upon their upper surface for the spinal nerves, and perforated by a round hole at their basis for the vertebral artery, the direction of which is upwards. Each transverse process has two roots; the posterior springs from between the articular processes, at the junction of the pedicle with the arch, as is the case with the transverse processes in the dorsal and lumbar regions: the anterior arises from the side of the body of the vertebra.

6th. The articular surfaces are oval in shape, the superior being convex, and directed obliquely backwards and upwards; the inferior being concave, and directed obliquely forwards and downwards.

7th. The notches, nearly of equal size, are small, and anterior to the articular processes, as in the other vertebræ; but are behind the anterior root of the transverse processes.

8th. The vertebral foramen is large and triangular.

DEVIATIONS.

The first, or atlas, consists of a large bony ring enclosing an irregular hole, which in the fresh state is divided into two parts by the transverse ligament; the anterior being occupied by the odontoid process of the axis, the posterior by the spinal chord. Instead of a "body," it presents anteriorly a small arch of bone, marked by a tubercle in front, and a smooth oval surface behind, which articulates with the odontoid process of the axis. At the extremities of this anterior arch the bone acquires great density and thickness, and presents upon its upper and lower aspects the articular processes, the superior of which, horizontal and oval from before backwards, look upwards and inwards, and articulate with the condyles of the occipital bone; the inferior, circular, slightly oval, and directed downwards and inwards, articulate with the second cervical vertebra. The parts of the vertebra which support the articular processes are called the "lateral masses," and are marked internally by rough surfaces, which afford attachment to the transverse ligament. The transverse processes are long, and terminate in a tubercle; they are pierced at their basis by the foramen for the vertebral artery, the direction of which is upwards and backwards; behind the superior articular surface is a groove, which marks the continued course of this vessel. The spinous process is represented by a small tubercle upon the middle of the posterior segment of the ring, which is considerably larger than the anterior.

The second, or axis, is distinguished by its large tooth-like, or odontoid, process. This process presents anteriorly an oval surface for articulation with the ring of the atlas, and posteriorly a smooth surface, which moves against the transverse ligament, whilst its apex presents an acuminated top, to which the check ligaments are attached. The laminæ are very thick and strong, and terminate behind in the spinous process, which is strong and bifid: the vertebral foramen is heart-shaped, the apex being behind. The superior articular surfaces are large, and directed a little outwards, whilst the inferior, looking downwards and forwards, are smaller and flat.

The transverse processes, not bifid, are short and directed downwards, the aspect of the foramen for the vertebral artery being directed obliquely upwards and outwards; the superior notches are behind the superior articular processes, whilst the inferior notches are before the inferior processes.

The seventh cervical vertebra is called vertebra prominens, from the length of the spinous process, which projects backwards and terminates in a tubercle; the foramina, when they exist in the transverse processes, give passage to the vertebral veins: the transverse processes are often of such length as to resemble rudimentary ribs.

In examining the peculiar characters of the different vertebræ, it is best to select one from near the centre of each class: thus the 3d lumbar, the 6th or 7th dorsal, and the 4th or 5th cervical, offer the best examples of the class to which each belongs; for as the cervical vertebræ approach the dorsal, they begin to assume more or less the characters of the latter, and the last dorsal vertebra, upon its under surface, presents the characters of a lumbar vertebra. But there is no difficulty in determining the region from which any vertebra comes, however faintly its characters may be marked. The transverse processes of all the cervical vertebræ are perforated by a foramen which transmits the vertebral artery or vein; the dorsal vertebræ present upon the bodies a smooth surface of articulation for the head of the rib: the lumbar vertebræ are distinguished by the absence of these characters.

FALSE VERTEBRÆ,

eight or nine in number: the five superior form the sacrum, the four inferior the coccyx.

Os Sacrum.

Figure, triangular, *the base* resembling a lumbar vertebra; *the apex* presenting a small oval surface to articulate with the os coccyx; *the sides* presenting two surfaces: the superior, large and irregular for articulation with the ilium; the inferior thin for the attachment of the sacrosciatic ligaments.

Pelvic surface, anterior, smooth, concave from above downwards, traversed by four transverse lines, and presenting on either side of the median lines four holes, called *anterior sacral*, for the transmission of the anterior sacral nerves. The projection which the sacrum forms with the last lumbar vertebra is called " the promontory."

Dorsal surface, irregularly convex, rough, presenting in the median line irregular processes of bone (spinous processes), and more externally on either side tubercles of bone analogous to the articular processes of the true vertebræ. On either side of the median line are the *posterior sacral foramina*, for the transmission of the posterior sacral nerves.

Vertebral canal, at the base large, and of triangular form, runs at the dorsal aspect of the bone, and terminates in a triangular fossa at its apex, where it is bounded on either side by two tubercles, which are in general prolonged to join the base of the os coccyx. It contains the cauda equina.

Os Coccyx.

Figure, triangular, the base articulating with the sacrum.

Anterior surface, smooth, and supports the extremity of the rectum.

Posterior surface, rough, for attachment of ligaments and muscles.

Cornua, are two, placed superiorly, which unite with the last tubercles of the sacrum.

The coccyx is usually composed of three or four pieces, of which the first, which articulates with the sacrum, is the largest.

THE THORAX

is formed by the dorsal vertebræ posteriorly, the ribs laterally, and the sternum anteriorly.

The Ribs,

twelve in number, are divided into *seven true* and *five false;* the two lowest of the false being called *abdominal,* or *floating ribs.*

The true ribs are each attached to the sternum by separate cartilages.

The three superior false ribs have their cartilages attached to each other and to the cartilage of the seventh rib.

The false or floating ribs have their cartilages free.

Common Characters of a Rib.

The head, round, and divided by a ridge into two articular surfaces, which are received into depressions on the sides of the bodies of the dorsal vertebræ, the ridge affording attachment to a ligature which connects the rib to the intervertebral fibro-cartilage.

The neck, narrow and round; at its union with the shaft is

The tubercle, a prominence of bone, with a smooth surface looking backwards to articulate with the transverse process of the vertebra beneath, and a rough surface for the attachment of the posterior costo-transverse ligament.

The angle, is marked by a rough line, and is the point where the rib makes its great turn to circumscribe the thorax.

The shaft, that portion of the rib which extends from the angle to its sternal end, presents an external smooth convex surface, and an inter-ual concave one; its upper edge being round and smooth, its lower edge thin, and grooved for the intercostal vessels, and its sternal extremity offering an oval pit for the reception of the costal cartilage.

DEVIATIONS.

First Rib,

Has no angle, is short and flat, the head articulating with only one vertebra, has no ridge, and one articulating service; its sternal extremity being thick and strong; one surface is directed upwards, the other downwards; the superior presenting a ridge for the attachment of the anterior scalenus muscle, which separates two grooves, one for the subclavian artery, the other for the subclavian vein.

The Second Rib,

Longer than the first, presents externally an oblique prominent ridge for the attachment of the first digitation of the serratus magnus muscle.

Eleventh and Twelfth Ribs,

have neither angle, tubercle, nor groove, are very short, and the head resembles that of the first.

THE STERNUM,

Consists of three pieces.

Figure, flat, elongated, broad above, narrower in the centre, and pointed inferiorly.

Anterior surface, is marked by four transverse lines, and is rather convex.

Posterior surface, smooth and concave.

Upper piece, quadrilateral and thick, is concave from side to side upon its upper edge, and presents at each superior angle two semilunar depressions for articulation with the clavicles; its lower edge is united to the second piece; its lateral edges receive on either side the cartilage of the first rib and half that of the second.

Middle piece, long and narrow; receives, by five depressions upon its lateral edges, the cartilages of the five inferior true ribs, and by a notch at its superior angle half the cartilage of the second rib. Its inferior extremity is long and thin, and ends in a cartilaginous epiphysis, called the

Ensiform cartilage, which is generally bifid, and pierced by a foramen, its direction being variable.

THE PELVIS,

Is formed of the os sacrum, os coccyx (both of which bones have been already described), and the two ossa innominata.

OS INNOMINATUM,

Consists in early life of three bones, the ilium, ischium, and pubes. We shall describe each of these bones separately.

OS ILIUM.

Situation, upwards and outwards in regard to the pelvis, forming the upper part of the acetabulum, and bounding the lower lateral part of the abdomen.

It forms the upper and outer part of the acetabulum, joining the pubes anteriorly, and the ischium posteriorly.

The *dorsum* is rough at the posterior fifth for the gluteus maximus; and presents two semicircular lines, of which the superior passes from the anterior superior spine to the sacro-sciatic notch, and the inferior from the anterior inferior spine towards the same notch: the gluteus minimus arises from between the curved lines: the gluteus medius from the space between the superior curved line and the rough surface for the gluteus maximus. In the centre is a nutritious foramen. The venter, which forms the iliac fossa, has also a nutritious foramen; there is a rough articular surface for connection with the side of the sacrum, and a small portion smooth and immediately above the sciatic notch, which enters into the formation of the true pelvis.

Processes.

The *crest*, which forms the upper border of the bone.

The *anterior superior spine*, which terminates the crest anteriorly.

The *anterior inferior spine*, immediately above the acetabulum.

Both spines are separated by a *notch*.

The *posterior superior spine*, which terminates the crest behind.

The *posterior inferior spine*, separated from the former by a small notch.

The *Ilio-pectineal eminence*, marks the union of this bone with the following,

OS PUBIS.

Situation, fore part of pelvis, and internal part of acetabulum.

Body, forms the internal and superior part of the acetabulum.

Horizontal ramus, passes inwards to meet the opposite ramus. *Descending ramus* passes vertically downwards from the inner extremity of the horizontal ramus to join the ramus of the ischium.

Symphysis pubis, is formed by the apposition of both descending rami.

The angle is formed by the junction of the horizontal and descending rami. From the angles, a rough line, the crista, about an inch in length, runs horizontally outwards, and terminates in a tubercle called the spine.

The obturator groove, for the obturator vessels and nerve, is situated upon the under surface of the horizontal ramus.

The obturator or thyroid foramen is the oval

space bounded by the pubes above, and by the ischium below.

THE ISCHIUM.

Situation, lower, outer and back part of the pelvis.

Body, forms the outer and back part of the acetabulum; immediately beneath this cavity is a groove for the tendon of the obturator externus muscle. The anterior thin edge of the body assists to form the obturator foramen, the posterior to form the sciatic notch.

Spine, projects backwards and inwards, and gives attachment to the lesser sacro-sciatic ligament.

Tuberosity, the thickest part of the bone, on which we rest when sitting: gives attachment to the greater sacro-sciatic ligament. Between this process and the spine is a pulley-like surface, for the tendon of the obturator internus muscle.

Ascending ramus, turns forwards, upwards. and inwards, and joins the descending ramus of the pubes; it bounds the obturator foramen by its outer thin edge, and the lower aperture of the pelvis by its thick one.

ACETABULUM.

Is formed by the union of the ilium, ischium, and pubes; the ilium forming less than two-fifths; the ischium more than two-fifths; and the pubes the smallest part. It presents an articular surface for the head of the femur, and is surrounded, in the greater part of its circumference, by a margin, which is prominent at the upper and outer part; but deficient at the lower and inner part, towards the obturator foramen. The part where the margin is wanting is called the cotyloid notch; to its borders is attached the ligamentum teres of the hip joint.

SUPERIOR EXTREMITY,

Consists of the shoulder, the arm, the forearm, and the hand.

THE BONES OF THE SHOULDER,

Are the clavicle and the scapula.

THE CLAVICLE.

Situation, from the notch in the upper piece of the sternum, to the acromion process of the scapula.

Figure, curved like the italic *f*.

Sternal end, thick, presenting a triangular articular surface; its edge is rough for ligaments.

Body, cylindrical towards sternum, flat and expanded towards its acromial end. Upon its under surface are, 1, towards the sternal extremity, a rough surface for the costo-clavicular ligament; 2, towards the acromial extremity, a rough line for the coraco-clavicular ligaments, and between both a groove for the subclavius muscle, in which groove is found the nutritious foramen.

Acromial end, rough and flattened, passes over the coracoid process to meet the acromion, with which it articulates by a small oval surface.

SCAPULA.

Situation, upper and back part of thorax, extending from the second to the seventh rib.

Figure, triangular.

Costæ or edges.—The superior or cervical is the shortest, and is interrupted by a notch for the suprascapular nerve; the inferior or axillary is next in size, and is the thickest; the posterior or vertebral is the longest, and is also called the base of the scapula.

Angles.—The superior posterior angle is acute and prominent, the inferior angle is thicker and rounded, and the anterior angle has connected to it the neck of the bone.

Costal surface, or *subscapular fossa*, is slightly concave and divided by three or four lines, which run from above obliquely downwards and inwards.

Dorsal surface is divided unequally, by the spine, which thus gives rise to the spinous fossæ.

Spine arises by a small triangular surface, at the vertebral margin, and proceeds forwards, becoming more elevated, and terminates in the acromion, which surmounts the shoulder joint and articulates with the acromial end of the clavicle by a small oval surface.

Supraspinous fossa, situated above the spine, is deep, and presents a nutritious foramen; it lodges the supraspinatus muscle.

Infraspinous fossa, larger, irregularly concave and convex, affords attachment to the infraspinatus muscle, the teres minor muscle, and by an inferior rough surface, to the teres major muscle.

Coracoid process overhangs the inner and upper part of the glenoid cavity. This process has a crooked appearance, and gives attachment to the pectoralis minor, the short head of the biceps and coraco-brachialis muscles, also to ligaments.

Glenoid cavity, articulates with the head of the humerus, is shallow and oval, being broader below, and giving attachment by its upper narrow part to the long head of the biceps muscle.

Neck is the contracted portion of the scapula immediately behind the glenoid cavity; it gives attachment to the capsular ligament of the joint.

OS HUMERI.

Articulates with the scapula above and radius and ulna below.

Head forms a small section of a large sphere, is smooth and covered with cartilage in the recent state. Its direction is obliquely backwards and inwards to the glenoid cavity of the scapula.

Neck, a slight contracted line, rough for the attachment of the capsular ligament. It is united to the shaft at an obtuse angle.

Tuberosities exist at the junction of the shaft with the neck. The largest is posterior, and affords attachments by three flat surfaces to the supra-spinatus, infra-spinatus, and teres minor muscles; the smallest being anterior, is more prominent, and gives attachment to the subscapular muscle.

Bicipital groove is between both tuberosities, lodges the long biceps tendon, and affords attachment by its anterior margin to the pectoralis major muscle, and by its posterior margin to the latissimus dorsi and teres major muscles.

Shaft is rather twisted, the upper extremity cylindrical, the lower flattened; two longitudinal ridges, arising one from the outer, the other from the inner condyle, and affording attachment to intermuscular septa, divide it into an anterior and a posterior region. The external longitudinal ridge is interrupted by an oblique groove which indicates the course of the muscular-spiral nerve and vessels: upon the external and central aspect is a rough portion for insertion of the deltoid muscle, and upon the inner side of the bicipital groove is a slight rough line for the insertion of the coraco-brachialis muscle. The nutritious foramen is directed downwards towards the elbow joint.

Internal condyle is prominent and sharp, affording attachment to the pronator and flexor muscles and to the internal lateral ligament.

External condyle is less prominent, but descends nearer the elbow joint, and gives attachment to the supinator and extensor muscles, and to the external lateral ligament.

Articulating surfaces. First, a small round *head*, situated externally and nearer the anterior than the posterior part of the bone, for the radius. Second, the *trochlea* or pulley-like surface placed internally for the ulna; being much below the level of the *head*, so as to give the articular surface an oblique direction from above downwards and inwards.

Small sigmoid fossa at the fore part of the trochlea to receive the coronoid process in the bent position of the forearm.

Great sigmoid fossa at the back part of the trochlea, to receive the olecranon process in the extended position of the forearm.

ULNA.

Situation at the inner side of the forearm.

Upper extremity articulates with the humerus and radius.

Olecranon process, forming the projection of the elbow, is the highest point of the bone, and by its extremity gives attachment to the triceps muscle; beneath this it is smooth for a bursa mucosa.

Coronoid process, anterior to the olecranon, is

smaller than it, and gives attachment to the brachialis anticus muscle, some of the flexors and pronators of the forearm, and to the internal lateral ligament.

Lesser sigmoid cavity, oval, receives the side of the head of the radius.

Greater sigmoid cavity has its long axis from before backwards, is divided in the centre by a transverse line, and by a vertical ridge, into two lateral portions, is bounded above by the olecranon process, below by the coronoid process, and receives the articular extremity of the humerus.

Shaft presents three surfaces and three edges; the anterior surface affords attachment to the flexor digitorum profundus muscle, and is pierced by the nutritious foramen, which is directed upwards. The posterior gives attachment to the anconeus muscle and to the extensor muscles of the thumb; the internal is subcutaneous for its greatest extent. Of the edges, that directed towards the radius is the best marked; it gives attachment to the interosseous ligament.

Carpal extremity slender and rounded, presents the *head*, which by its side articulates with the radius, and by its carpal aspect with the fibro-cartilage of the wrist-joint; also the *styloid process* at the inner margin, which by its apex gives attachment to the internal lateral ligament of the carpus, and by a depression at its root, to the fibro-cartilage; posteriorly it presents a groove for the tendon of the flexor carpi ulnaris muscle.

RADIUS.

Situation, external to the ulna; it is also shorter than that bone, by the length of the olecranon.

Head, a superficial circular cavity, articulating above with the humerus, and latterly with the lesser sigmoid cavity of the ulna.

Neck, an inch in length, narrow and rounded, and terminating in the

Tubercle, a prominent process into which the tendon of the biceps muscle is inserted.

Shaft, triangular, by its anterior surface affording attachment to the flexor longus pollicis and pronator quadratus muscles, by its posterior convex surface to the supinator radii brevis and extensor muscles of the thumb; its external surface being round and convex, and rough near its centre for the pronator radii teres muscle. The inner edge of the shaft is sharp for the attachment of the interosseous ligament, and the nutritious foramen upon the anterior surface of the shaft is directed upwards.

Carpal extremity expanded, gives attachment by its anterior edge to the anterior carpal ligament; upon its posterior edge to a shallow groove at its ulnar side for the extensor communis digitorum and indicator muscles, a second more externally, narrow and oblique for the tendon of the extensor secundi internodii pollicis muscle, and a third more external than the last for the tendons of the extensores carpi radialis longior and brevior. Upon the radial edge of this extremity is the mark of the passage of the extensor ossis metacarpi and the extensor primi internodii pollicis muscles.

Styloid process, prolonged on the outer side of the bone downwards, giving attachment to the external lateral ligament of the carpus.

A cavity on the inner border of the carpal extremity for articulation with the ulna.

Carpal aspect presents two articular surfaces, divided by a slight elevation; the external being triangular for articulation with the os scaphoides, the internal square for the os lunare.

CARPUS.

This part consists of two rows of small bones, four bones in each row, placed between the forearm and metacarpus. It is convex and rough upon its dorsal aspect, and concave upon its palmar aspect, where the vessels, nerves, and tendons of the flexor muscles are situated; towards the radius and interarticular cartilage of the wrist, it is convex; towards the metacarpus it presents articular surfaces for the metacarpal bones.

FIRST ROW.

Os Scaphoides.

Situation, on outer and upper part of the carpus, next the styloid process of the radius.

Articulations, with the radius above, with

trapezium and trapezoides below, and with the lunare and magnum internally.

Os Lunare.

Situation, between the scaphoides and cuneiforme.

Articulations, with the radius above, with the magnum and unciforme below, with the scaphoid externally, and with the cuneiforme internally.

Os Cuneiforme.

Situation, at the internal and upper part of the carpus between the lunare and pisiforme.

Articulations, with the lunare externally, with the carpal fibro-cartilage above, with the unciforme below, and upon its palmar aspect with the pisiforme.

Os Pisiforme.

Situation, upon the inner and palmar aspect of the cuneiforme.

Articulation, by a circular small surface with the cuneiforme.

SECOND ROW.

Os Trapezium.

Situation, between the scaphoides and metacarpal bone of the thumb.

Articulations, with the scaphoides above, with the metacarpal bone of the thumb below, and internally with the trapezoides and side of the metacarpal bone of the index finger.

Groove, upon its palmar aspect, for lodging the tendon of the flexor carpi radialis muscle. The surface which articulates with the first metacarpal bone is concave in one direction, convex in the other.

Os Trapezoides.

Situation, between the trapezium, magnum, scaphoides, and metacarpal bone of the index finger.

Articulations, with the scaphoides above, the trapezium externally, the magnum internally, and below with the metacarpal bone of the index finger.

Os Magnum.

Processes, head, neck, and body.

Situation, between the scaphoides and lunare, and the second, third, and fourth metacarpal bones.

Articulations, by its head with the scaphoides and lunare above, by its base with the second, third, and fourth metacarpal bones below, externally with the trapezoides, and internally with the unciforme.

Os Unciforme.

Situation, between the cuneiforme and metacarpal bones of the ring and little fingers.

Articulations, with the lunare above, the magnum externally, the cuneiforme internally, and below with the fourth and fifth metacarpal bones.

Process, a hook-like process upon its palmar aspect.

The carpal bones in their natural state are so arranged that the os magnum and os unciforme form a convex head which is received into a concavity formed by the ossa scaphoides, lunare, and cuneiforme. The trapezium and trapezoides form a concavity which receives the convex surface of the os scaphoides.

METACARPUS.

This part consists of five long bones placed between the carpus and the fingers.

Bases, irregularly flattened for articulation with the second row of carpal bones, with rough surfaces for the attachment of ligaments.

Heads, convex, and rounded for articulation with the first bones of the fingers, and affording attachments for the capsular, transverse, and lateral ligaments.

Bodies.—The *first*, for the thumb, is the thickest and shortest; its carpal articulating surface corresponds with the saddle-like articulating surface of the os trapezium; the *second*, *third*, and *fourth* present on the dorsal aspect of each a prominent line, which bifurcates towards the head; the *fifth* presents an oblique line on its dorsal aspect, which divides it into two surfaces. The third is the longest.

FINGERS.

The fingers are composed of three phalanges, except the thumb, which has only two.

The Metacarpal, or first Phalanges, are five in number, their bases presenting oval concavities for the heads of the metacarpal bones; their anterior extremities are convex from before backwards, and concave from side to side, to articulate with the second phalanges.

The Middle, or second Phalanges, are four in number, and smaller than the first: their bases present pulley-like surfaces, to form a ginglymoid joint with the first phalanges, and at their anterior extremities resemble the first.

The Ungual, or third Phalanges, are five in number, and the smallest. By their bases they form a ginglymoid joint with the middle phalanges; and their extremities are convex upon their dorsal aspects for the support of the nail, whilst their palmar aspects are irregularly tuberculated for the extremities of the fingers.

The Sesamoid Bones are sometimes absent. In general two are to be found between the metacarpal bone of the thumb and its first phalanx, and one or two at the corresponding joint of the index finger.

LOWER EXTREMITY.

The bones proper to the lower extremity are

the femur, the tibia, the fibula, the patella, seven tarsal bones, five metatarsal bones, and fourteen phalanges.

THE FEMUR.

Head, forms nearly two-thirds of a sphere, and is received into the acetabulum, being directed forwards, upwards, and inwards; below its centre is a rough pit for the ligamentum teres, and its junction with the neck is marked by a rough line. With the exception of these parts it is covered by cartilage in the recent state.

Neck, extends obliquely downwards and outwards from the head to the shaft, which it joins at an obtuse angle; it is flattened anteriorly and posteriorly, and its lower edge is much longer than its upper.

Great trochanter, is continued upwards from the shaft, but does not extend as high as the head, is situated externally, and gives attachment to the gluteus medius, gluteus minimus, and pyriformis muscles.

Digital fossa, is situated at the root of the great trochanter, and affords insertion to the external rotator muscles.

Lesser trochanter, is situated posterior and internal to the shaft, and has inserted into it the common tendon of the psoas and iliac muscles.

Intertrochanteric line, as the name implies, passes (obliquely) from one trochanter to the other.

Shaft, is broad at either extremity, particularly towards the knee, and is narrow and triangular in the centre; it is arched and smooth anteriorly, where it affords attachment to the crureus muscle; at its posterior part it is concave, and presents a rough line called linea aspera, about the upper third of which, and directed upwards, is the foramen for the nutritious vessels.

Linea aspera, for the attachment of muscles, is best marked for the central third of the shaft; towards the upper third it bifurcates, one ridge running to each trochanter; inferiorly it also divides into two ridges, which pass to either condyle, the inner one being interrupted where the popliteal vessels pass over it. By these inferior ridges a flat triangular surface of the femur is bounded laterally, which receives the name of popliteal surface.

External condyle, is larger and projects more upon the anterior surface of the femur than the internal: its articulating surface is broader and ascends higher upon the shaft.

Internal condyle, narrower than the external; it also descends lower, in order that both condyles should rest upon the tibia in the natural oblique direction of the femur.

External tuberosity, above the external condyle for the attachment of the external lateral ligament of the knee-joint.

Internal tuberosity, above the internal condyle and more prominent than the external, for the internal lateral ligament and the insertion of the great adductor tendon.

Trochlea, between the condyles on their anterior aspects, supports the patella when the leg is extended.

Intercondyloid fossa, is situated posteriorly between the condyles.

PATELLA.

Shape, triangular, the base superiorly having the extensor muscles inserted into it, the apex below, to which is attached the ligamentum patellæ.

External surface, presents a fibrous appearance, and is convex.

Internal surface, presents two articular surfaces, divided by a prominent edge: the internal surface is the deepest, whilst the external is broad and shallow.

TIBIA.

Shaft, is triangular, presenting three surfaces, separated by corresponding edges. The inner surface is subcutaneous, except at its upper part, where the tendons of the sartorius, gracilis and semitendinosus muscles pass over it: it terminates in the inner malleolus; the external surface is covered by the belly of the tibialis anticus muscle for its upper two-thirds, and is here concave; inferiorly the bone is flattened to support the tendons of the extensor muscles of the toes;

the posterior surface presents an oblique line directed upwards and outwards for the attachment of the soleus, popliteus, and deep flexor muscles; the nutritious foramen, which is the largest in the body, is a little below this line, and directed downwards. Of the edges, the anterior is most prominent and subcutaneous; the inner is less defined, and the outer is well marked, affording attachment to the interosseous ligament.

Upper or femoral extremity, which articulates with the femur, is expanded from side to side; anteriorly its circumference is convex; posteriorly it is slightly hollowed. Two concave surfaces, condyles, of which the inner is the deeper, receive the condyles of the femur. Between the condyles is the spine, a bifid eminence, before and behind which are depressions for the insertion of the crucial ligaments and the semilunar cartilages.

Tuberosities, are three in number, two lateral, and one anterior; the anterior is most prominent for the insertion of the ligament of the patella; the inner is less so, and affords attachment to the internal lateral ligament of the knee-joint, and the semimembranosus muscle; the outer is the least developed, and has an articulating surface for the fibula.

Lower or tarsal extremity, is much smaller than the upper and quadrilateral, the anterior edge being convex for the passage of the extensor tendons: the posterior, marked by a groove, for the passage of the tendon of the flexor longus pollicis muscle; the external presenting a rough triangular surface for the fibula, and the internal terminating in the internal malleolus.

Internal malleolus, is convex and subcutaneous internally; its articular aspect being smooth, to articulate with the astragalus; it gives attachment by its inferior border to the internal lateral ligament of the ankle-joint, and is grooved posteriorly for the tendons of the tibialis posticus and flexor longus digitorum muscles.

Inferior articular surface, rests upon the crown of the astragalus; it is quadrilateral and concave from before backwards, and joins the small articular surface of the internal malleolus at a right angle.

FIBULA.

Situation, on the outer side of the tibia.

Shaft, triangular and twisted, gives attachment to the peronei muscles by its external surface, to the soleus and flexor pollicis muscles by its posterior surface, and to the extensor muscles by its anterior surface. Its inner edge is well developed to afford attachment to the interosseous ligament, and the nutritious foramen, directed downwards, is placed upon the posterior aspects of the bone.

Head, articulates with the tibia by a small circular surface directed upwards and inwards, and affords attachment to the external lateral ligament of the knee-joint, and the biceps flexor muscle by a rough projection situated posteriorly.

Neck, is the small constricted part by which the head is united to the shaft.

Tarsal extremity, presents a large oval process called the external malleolus, about two inches and a half above which the shaft is constricted.

External malleolus, is larger than the internal, on a line posterior to which it is situated; by its edge it affords attachments to the external lateral ligaments of the ankle-joint. Its outer surface is convex and subcutaneous, the inner or articular being smooth to articulate with the outer side of the astragalus; above this surface is a triangular rough aspect for the inferior connection of the bone with the tibia.

TARSUS.

Forms the posterior part of the foot; the bones which compose it are seven in number, and articulate with each other. It is connected above to the tibia and fibula, below it is hollow, and anteriorly it unites by articular surfaces with the five metatarsal bones. The seven tarsal bones are the calcaneum, astragalus, navicular or scaphoid, cuboid, and three cuneiform.

CALCANEUM, OR OS CALCIS.

Situation, at the posterior and under part of the tarsus.

Posterior aspect, elongated to form the heel,

presents a smooth surface above for a bursa mucosa, and a rough surface below for the insertion of the tendo-achillis.

Anterior aspect, articulates by a smooth surface with the cuboid bone.

Superior aspect presents two articular surfaces, which support the astragalus; they are separated by a groove, into which a strong interosseous ligament is inserted.

Inferior aspect is irregular, and presents two small tubercles, and two lines for the attachment of muscles and ligaments.

External aspect presents a small tubercle for the insertion of the middle external lateral ligament of the ankle-joint; it is also slightly grooved for the passage of the peronei tendons.

Internal aspect presents a projection for the internal lateral ligament, and a fossa for the lodgment of posterior tibial vessels and nerve and tendons of muscles.

ASTRAGALUS.

Situation, at the upper and middle part of the tarsus, between the malleoli at either side : the tibia above, and the os calcis below.

Crown, broader anteriorly than posteriorly, presents three articular surfaces; one, large superiorly, to articulate with the tibia; the others, one on either side, to articulate with the malleoli.

Head, convex and smooth, is directed forwards and inwards, and is larger than the concavity of the navicular bone with which it articulates.

Neck, the contracted portion between the crown and the head is rough for the attachment of ligaments.

Inferior aspect presents two articular surfaces for connexion with the os calcis: they are separated by a groove, from which ligaments pass to the groove in the os calcis.

Posterior aspect is narrow, and presents a groove for the tendon of the flexor longus pollicis muscle, and a pointed eminence for the posterior external lateral ligament of the ankle-joint.

NAVICULARE.

Situation, middle of the tarsus.

Posterior aspect is smooth and concave, to articulate with the head of the astragalus.

Anterior aspect presents two vertical lines, which divide it into three smooth surfaces for articulation with the three cuneiform bones, and, in general, a small articular surface, externally, where it touches the cuboid bone.

Tubercle, situated internally for the insertion of the tendon of the tibialis posticus muscle.

CUBOIDES.

Situation, outer and anterior part of the tarsus.

Upper surface, flat and rough for ligaments.

Lower surface, tubercular behind, for the calcaneo-cuboid ligament, and presenting a groove anteriorly for the tendon of the peroneus longus muscle.

Posterior surface, smooth and concave transversely, to articulate with the os calcis.

Anterior surface presents two articular surfaces; the external triangular to articulate with the metatarsal bone of the little toe, the inner square for the fourth metatarsal bone.

External surface narrow and free.

Internal surface presents posteriorly a small articular surface for the naviculare, and anteriorly another, which is larger, to articulate with the outer side of the external cuneiform bone.

INTERNAL CUNEIFORM BONE.

The largest of the three articulates behind with the navicular; before, with the first metatarsal bone; externally, with the middle cuneiform and side of the second metatarsal bone.

Tubercle is situated inferiorly for the insertion of the tendon of the tibialis anticus muscle, and part of the tendon of the tibialis posticus.

MIDDLE CUNEIFORM BONE.

The smallest, articulates behind with the navicular bone; before, with the second metatarsal bone; externally, with the external cuneiform; and internally, with the internal cuneiform.

EXTERNAL CUNEIFORM BONE.

Articulates behind with the navicular bone; before, with the third metatarsal bone; internally, with the middle cuneiform and side of the second metatarsal bone; and externally, with the cuboid and fourth metatarsal bone.

METATARSUS

is the middle part of the foot, and is composed of five long bones placed between the tarsus and the toes.

FIRST METATARSAL BONE,

the shortest and thickest, is convex above and concave below.

Posterior extremity is smooth and oval, to articulate with the internal cuneiform bone.

Anterior extremity, round, to articulate with the first phalanx of the great toe. The under surface of this extremity articulates with two sesamoid bones, and receives the insertion of the tendon of the peroneus longus.

SECOND METATARSAL BONE,

the longest.

Posterior extremity articulates with the three cuneiform bones, and also with the third metatarsal bone.

Anterior extremity presents a round head, for articulation of the second toe; it is separated from the shaft of the bone by a groove.

THIRD METATARSAL BONE.

Posterior extremity articulates with the third cuneiform bone.

Anterior extremity resembles the second, and articulates with the third toe.

FOURTH METATARSAL BONE.

Posterior extremity articulates with the cuboid bone, and by its inner side with the third cuneiform.

Anterior extremity resembles the second, and articulates with the fourth toe.

FIFTH METATARSAL BONE.

Posterior extremity articulates with the cuboid bone by a surface directed obliquely upwards and outwards.

Anterior extremity resembles the second, and articulates with the fifth toe.

PHALANGES, OR TOES,

are composed of fourteen bones, three to each toe, except the first, which has only two.

FIRST PHALANGES.

Shape, convex above, concave below, and the longest.

Anterior extremities, convex from above downwards, and concave laterally; form ginglymoid articulations with the second phalanges.

Posterior extremities are large, and present rounded concavities for the heads of the metatarsal bones.

SECOND PHALANGES.

Anterior extremities resemble those of the first phalanges.

Posterior extremities, concave from above downwards, and convex transversely, articulate with the first phalanges.

THIRD PHALANGES.

Anterior extremities, pyramidal in form, support the nails on their upper surface, and are rough on their under surface.

Posterior extremities resemble the posterior extremities of the middle phalanges.

SESAMOID BONES

are uncertain in development; two are generally found at the base of the first phalanx of the great toe, and one at that of the fifth toe. They are also frequently developed in the tendons which cross the sole of the foot in those situations where they are subjected to much pressure.

THE SKULL,

Consists of eight bones, four of which are symmetrical, viz., the frontal, the ethmoid, the sphenoid, and the occipital; and four arranged in pairs, viz., two parietal, and two temporal.

FRONTAL BONE.

Situation.—Upper and anterior part of the skull.

External, or frontal aspect, is convex and arched, and presents a median vertical depression, which marks the original division of the bone into two.

Nasal spine, at the inferior part of the median depression or line.

Frontal eminences, on either side of the median line, marking the points of ossification of the bone.

Superciliary arches, two convexities which extend for about an inch on either side of the median line inferiorly.

The glabella, or eminence of frontal sinuses, between, but a little above, the superciliary arches.

Orbital arches, form the upper edges of the orbits, and present towards their inner third the supraorbital holes or notches, for the passage of the frontal nerve and artery.

External angular processes, at the outer terminations of the orbital arches.

Internal angular processes, at the inner terminations of the orbital arches.

Temporal ridges, extend backwards and upwards from the external angular processes.

Internal or cerebral aspect, is concave, and presents in the median line a groove, which corresponds to the superior longitudinal sinus.

Crest, at the commencement of the median groove, gives attachment to the falx.

Foramen cæcum, at the root of the crest, gives passage to a vein which opens from the nose into the longitudinal sinus.

Eminences and depressions, on either side of the median line, commonly stated to correspond with the convolutions of the brain.

Orbito-ethmoidal aspect, is irregular, forming, on either side, part of the orbits, and in the median line part of the nose.

Ethmoidal notch, quadrilateral, articulates with the ethmoid bone by a serrated margin; its edges being cellular to unite with the ethmoidal cells.

Anterior and posterior ethmoidal foramina, along the margins of the ethmoidal notch, giving passage to the nasal twig of the ophthalmic nerve and the ethmoidal arteries.

Orbital processes, triangular, and on either side of the ethmoidal notch, form the roofs of the orbits by their smooth concave surface, and on their convex surface are marked by the convolutions of the brain. Towards the external angular process of each is the fossa for lodging the lachrymal gland, and at the nasal margin is a depression for the reflected tendon of the superior oblique muscle.

Frontal sinuses, which exist only in the adult, at the antero-inferior part of the bone, and between its two tables, open at either side of the nasal process.

It articulates with twelve bones, viz., the sphenoid, ethmoid, two parietal, two nasal bones, two ossa unguis, two superior maxillary, and two malar.

PARIETAL BONES.

Situation, upper and lateral parts of cranium; they are quadrilateral.

External surface, convex, and marked by a semicircular ridge, which is continued from the temporal ridge of the frontal bone.

Four edges. The upper is the longest, and joins its fellow,—the anterior joins the frontal bone,—the posterior is very irregular, and joins the occipital,—and the inferior is thick where it joins the mastoid process, but thin and semicircular where it is overlapped by the squamous portion of the temporal bone.

Four angles. The anterior inferior is long and curved, and joins the sphenoid bone; upon its cerebral aspect is a canal or a groove, for the trunk of the middle meningeal artery. The posterior inferior angle is grooved upon its cerebral aspect to lodge part of the lateral sinus. The superior angles are rather rounded.

Internal surface, concave, and marked by the convolutions of the brain and the ramifications of the middle meningeal artery, presents along its upper edge a shallow groove, which, with its fellow, accommodates the superior longitudinal sinus; external to this groove are depressions marking where the glandulæ Pacchioni externæ were situated. Each parietal articulates with five bones, viz., the frontal, sphenoid, occipital, temporal, and with its fellow of the opposite side.

OCCIPITAL BONE.

Situation, posterior and inferior part of cranium.

Tuberosity, near the centre of external surface.

Superior transverse ridge, leads from either side of the spine transversely outwards.

Inferior transverse ridge, midway between tuberosity and foramen magnum.

Spine, leads from tuberosity down to the foramen magnum.

Foramen magnum, larger internally than externally, is of oval form, and transmits the medulla spinalis, vertebral arteries, and spinal accessory nerves.

Basilar process, passes forwards and upwards to join the sphenoid bone, is rough inferiorly for the attachment of the pharynx; and upon its cerebral aspect is smooth and concave from side to side to support the pons Varolii and basilar artery, and marked laterally by two superficial grooves, which lodge the inferior petrosal sinuses.

Condyles, smooth and oblong, look downwards, outwards, and backwards; each presents a roughness internally for the check ligaments.

Anterior condyloid foramina, situated before the condyles for the passage of the ninth pair of nerves.

Posterior condyloid foramina, behind the condyles, for the passage of small veins. They are sometimes absent.

Jugular eminences, external to either condyle; they form with the temporal bone the foramen basis cranii posterius.

Crucial ridges, on the cerebral aspect, formed by a transverse and vertical line : they give attachment to the falx major by the upper half of the vertical line, in which is a groove for the termination of the longitudinal sinus, and to the falx minor or falx cerebelli by its lower half; the transverse ridge gives attachment to the tentorium, and also presents a groove on either side for the lateral sinuses.

Four fossæ, two superior for the posterior lobes of the cerebrum, and two inferior and deeper for the lateral lobes of the cerebellum.

Grooves for the termination of the lateral sinuses, on each side of the foramen magnum above the jugular eminences.

It articulates with six bones, viz., the two parietal, the two temporal, the sphenoid, and the atlas.

TEMPORAL BONE.

Situation, at the lateral, middle, and inferior part of the skull. It is divided into squamous, mastoid, and petrous portions.

Squamous portion, semicircular, flat, and thin, forming part of the side of the skull and of the temporal fossa.

Zygomatic process, arises by two roots, one

anterior bounding the front of the glenoid cavity, the other posterior, which is bifurcated; the process then passes forwards, and presents a serrated edge to unite with the malar bone.

Zygomatic tubercle, is situated at the union of the zygomatic roots, and gives attachment to the external lateral ligament of the lower jaw.

Glenoid cavity, transversely oval, deep anteriorly for the reception of the maxillary condyle, and shallow behind, where it lodges a portion of the parotid gland.

Glasserian fissure, crosses the glenoid cavity in a direction obliquely forwards and inwards. It transmits the chorda tympani nerve, the laxator tympani muscle, and lodges the processus gracilis of the malleus.

Auditory process, commencing by the external meatus, leads inwards and forwards to the membrana tympani, and presents externally a rugged edge for the attachment of the cartilage of the ear.

Internal surface of squamous plate, is marked by the convolutions of the brain and by blood-vessels.

Mastoid process, at the posterior and inferior aspect of the bone.

Mastoid grooves, internal to the mastoid process for posterior belly of the digastric muscle.

Mastoid foramen, posterior to mastoid process for the transmission of a vein.

Mastoid fossa, in the cerebral aspect, for the lateral sinus.

Mastoid cells, in the interior of mastoid process.

Petrous portion, containing in its interior the internal ear, extends forwards and inwards, and presents three aspects, being of a prismatic form.

Aqueduct of the cochlea, a minute foramen on the under surface, posterior to the styloid process.

Styloid process, long and tapering, descends obliquely forwards and inwards, gives origin to the styloid muscles and ligament.

Vaginal process, a plate of bone between the glenoid cavity and the carotid foramen.

Stylo-mastoid foramen, between the styloid and mastoid processes for the exit of the portio dura nerve: near the stylo-mastoid foramen is a minute aperture for the auricular branch of the pneumo-gastric nerve.

Carotid canal, commences in front of the styloid process, turns forward, upwards, and inwards, and terminates by the side of the body of the sphenoid bone.

Jugular fossa, a smooth surface behind and a little external to the carotid foramen: it forms with the occipital bone the foramen lacerum posterius, which transmits the eighth pair of nerves and the jugular vein. In the bony septum, between the jugular fossa and the carotid canal, is a small foramen, which transmits the nerve of Jacobson to the tympanum.

Processus cochleariformis, a thin plate of bone

separating two canals, which are situated in the angle between the petrous and squamous portions; the superior of which transmits the tensor tympani muscle; the inferior forming the bony part of the Eustachian tube.

Superior petrosal ridge, separates the anterior from the posterior surface; it gives attachment to the tentorium, and is grooved for the superior petrosal sinus.

Depression for Gasserian ganglion, on the anterior extremity of the superior surface.

Hiatus Fallopii, a foramen on the superior surface for the passage of the Vidian nerve.

Meatus auditorius internus, on the posterior surface for the transmission of the seventh pair of nerves and a small artery

Aqueduct of the vestibule, a small slit-like opening posterior to the meatus internus.

The superior semicircular canal, forms an eminence on the superior petrosal surface.

The temporal articulates with five bones, viz., the parietal, malar, inferior maxillary, sphenoid, and occipital. It is connected also to the os hyoides by the stylo-hyoid ligament.

ETHMOID BONE.

Situation, in the ethmoidal notch of the frontal bone.

Crista galli, an angular process which divides the cerebral aspect into two equal portions, to which is attached the commencement of the falx cerebri.

Cribriform plate, on either side of the crista galli, concave, to lodge the olfactory bulbs, and perforated with holes, for the passage of the filaments of the olfactory nerves, the nasal division of the ophthalmic nerves, and some small blood-vessels.

Nasal plate or lamella, descends from the under surface of the crista galli, joins the sphenoid bone posteriorly, the vomer and nasal cartilage inferiorly, and the os frontis and nasal bones anteriorly.

Orbital plate, or Os planum, a smooth square plate of bone situated externally, and forming part of the orbit; in its upper edge are two notches, which, with those in the frontal bone, form the anterior and posterior ethmoidal foramina.

Ethmoidal cells, between the orbital plate and the nasal plate, being ten or twelve in number. They are divided into anterior and posterior. The former, the most numerous, communicate with the middle meatus; the latter, few and small in size, open into the superior meatus of the nose.

Superior spongy bone, descends in a curved manner outwards from the upper and posterior part of the bone, forming a short channel, called the superior meatus of the nose.

Middle spongy bone, larger and more curved, descends outwards, forming the middle meatus of the nose.

Infundibulum, a smooth groove leading from

the anterior ethmoidal cells to the middle meatus.

The ethmoid articulates with thirteen bones, viz., sphenoid, frontal, vomer, two nasal, two superior maxillary, two ossa unguis, two palatal, and two inferior spongy bones.

SPHENOID BONE.

Body, corresponds to the median line, and presents six aspects.

Posterior aspect, is rough for connexion with the basilar process of occipital bone.

Anterior aspect, presents the openings of the sphenoidal sinus.

Inferior aspect, presents the *rostrum or azygos process*, which articulates with the vomer.

Superior or cerebral aspect, presents a fossa, called sella turcica.

Lateral aspects, join the great alæ or wings.

Sella turcica, a deep fossa on the cerebral aspect of the bone, for lodging the pituitary body.

Posterior clinoid processes, two rounded eminences posterior to the sella turcica.

Anterior clinoid processes, two rounded eminences situated external and anterior to the sella turcica.

Olivary process, an eminence between the anterior clinoid processes, and anterior to the sella turcica, on which the optic commissure rests.

Carotid grooves, one on either side of the sella turcica for the internal carotid arteries.

Processes of Ingrassias or lesser wings, two thin plates of bone extending forwards and outwards from the anterior clinoid processes, presenting anteriorly a spine in the median line, to unite with the ethmoid bone, and a serrated margin on either side to unite with the frontal bone. Their posterior free edges afford attachment to the sphenoidal folds of the dura mater. The external extremity of each terminates in an acute point, their cerebral aspect being smooth, to support the anterior lobes of the brain.

Great wings, extend laterally from the sides of the body; each presents three aspects, one anterior and smooth to assist in forming the outer part of the orbit; the second posterior, concave, and marked by cerebral convolutions, to assist in forming the middle cranial fossa; and the third external, being divided by a crest; the portion above which forms part of the temporal fossa, and the portion below, part of the zygomatic fossa.

Spinous processes, extend backwards and outwards from the posterior termination of each great wing.

Pterygoid processes, two on either side, descend from the angle of junction of each great ala and the body. The external is broad and irregular; the internal is long and narrow, and terminates in a hook called the hamular process.

Pterygoid fossa, is the hollow between the pterygoid processes posteriorly.

Sphenoidal fissures or foramina lacera, on either side of the sella turcica, are placed between the greater and lesser wings; each gives transmission to the third, fourth, first branch of the fifth, and the sixth nerves, and to the ophthalmic vein.

Foramina optica, at the roots of the lesser wings, give passage to the optic nerves and ophthalmic arteries.

Foramina rotunda, posterior, and a little external to the bases of the foramina lacera, give passage to the second division of the fifth nerve.

Foramina ovalia, near the posterior terminations of the great wings, give passage to the third division of the fifth nerve.

Foramina spinosa, in the spinous processes, give passage to the middle meningeal artery of either side.

Foramina pterygoidea, at the roots of the pterygoid processes, for the passage of the Vidian nerves.

The sphenoid articulates with all the bones of the cranium, and with several of those of the face, viz., the vomer, two malar, and two palate bones.

BONES OF THE FACE.

MALAR BONE.

Situation, at the outer and under part of the orbit, forming the cheek.

Internal surface, convex, and of irregular quadrilateral form.

External angular process, at the upper and outer edge, joins the frontal bone.

Maxillary process, serrated, and rests at inner edge on the superior maxillary bone.

Zygomatic process, passes backwards, and supports the zygomatic process of the temporal bone by a serrated edge.

Upper edge, forms the outer and inferior margin of the orbit.

Orbital process, a thin plate of bone, which passes from the upper edge backwards and inwards.

Lower edge, thick and uneven for the attachment of the masseter muscle.

Temporal aspect, behind the zygomatic process, is smooth for the lodgment of the temporal muscle.

Foramina, two or three in number on its cutaneous aspect, for vessels and branches of lachrymal nerves. It articulates with the frontal, superior maxillary, temporal, and sphenoid bones.

SUPERIOR MAXILLARY BONE.

Body, quadrilateral, presents anteriorly the canine fossa for the levator anguli oris muscle, a slight depression above the sockets of the incisor teeth, the myrtiform fossa, for the depressor alæ nasi muscle, and the opening of the infra-orbital canal for the superior maxillary nerve. Inferiorly it is bounded by the alveolar border.

Malar process, external and superior, presents a rough surface which supports the malar bone.

Nasal process, internal and superior, is serrated above to join the frontal bone, and presents a groove to support the nasal bone. Its cutaneous aspect is perforated by two or three small holes for blood-vessels, its posterior aspect is deeply grooved to assist in forming the fossa for the lodgment of the lachrymal sac and duct, and its internal or nasal aspect is divided by a ridge, which supports the inferior spongy bone.

Orbital plate, triangular, assists to form the floor of the orbit.

Infra-orbital canal, passes from behind forwards between the plates of the orbital process, and terminates in the *infra-orbital foramen*, which transmits the infra-orbital nerve. It gives off a smaller canal, anterior dental, which conducts a small nerve to the incisor teeth.

Temporal aspect, presents behind the malar process a smooth depression for the temporal muscle, and more posteriorly and inferiorly a *tuberosity* which corresponds to the dens sapientiæ.

Posterior dental foramina are three or four small holes which give passage to the posterior dental nerves, and are found near the tuberosity.

Palatine process is thick internally, and rough where it joins its fellow; its circumference corresponds to the alveolar process, and its posterior edge is thin and serrated where it joins the palate bone; its upper surface is smooth and concave from side to side, forming part of the floor of the nose, and its under surface is rough, forming, with the gum, part of the hard palate.

Nasal Crest, at the union of the palatine processes, projects upwards to receive the vomer.

Nasal spine projects forwards, above and between the central incisor teeth.

Foramen incisivum, is common to both bones, and exists inferiorly at the anterior union of their palatine plates; superiorly it bifurcates and opens by two foramina, one to each nostril.

Antrum maxillare or *Antrum Highmori*, a large cavity in the body of the bone, bounded above by the orbital plate, below by the alveoli of the molar teeth, anteriorly by the canine fossa, and posteriorly by the temporal aspect; externally it corresponds to the malar process, and internally presents a large irregular opening.

The superior maxilla articulates with its fellow of the opposite side, with the frontal os nasi, os unguis, ethmoid, with the malar, vomer, inferior spongy bone, palate, and occasionally with the sphenoid.

PALATE BONE.

Horizontal or palate plate, quadrilateral, is concave and smooth above, and completes the floor of the nose, and rough below, where it completes the hard palate. Its anterior edge is serrated to join the superior maxillary bone, its posterior edge is concave and gives attachment to the soft palate, its inner edge is serrated where it joins its fellow, and sends a *crest* upwards to support the vomer; its outer edge joins the nasal plate.

Nasal spine projects backwards from the union of the palate plates.

Nasal process, broad and thin; its inner surface is divided into two by a ridge which supports the inferior spongy bone: above and below this ridge the process is slightly concave, to assist in forming the inferior and middle meatuses: its external surface is rough and marked by the posterior palatine vessels and nerves; its anterior thin edge assists to close the antrum, and the posterior edge joins the pterygoid processes of sphenoid bone.

Pterygoid process, the thickest part of the bone, of a wedge shape, inclines backwards and outwards; it presents three grooves, a central one, smooth, which completes the pterygoid fossa, and a rough one on either side to articulate with the extremities of the pterygoid processes.

Palatine foramina, at the junction of pterygoid and palatine processes, give passage to the posterior palatine nerve and vessels.

Orbital plate surmounting the nasal process, is divided into two by a notch, which forms with the sphenoid bone the *spheno-palatine foramen*.

Orbital process, is large and hollow, and of triangular form; it presents five surfaces, three articulated, two free. The anterior articulates with the superior maxilla; the posterior with the sphenoid; the internal with the ethmoid. The superior forms part of the floor of the orbit: the external looks into the zygomatic fossa.

Sphenoidal process, articulates with the body of the sphenoid bone, and is also hollow, forming the pterygo-palatine canal.

It articulates with six bones, viz., the corresponding palate bone, the superior maxilla, ethmoid, sphenoid, vomer, and inferior spongy bone.

INFERIOR SPONGY BONE.

Rough and convex towards the septum of the nose, and concave externally; presents a free margin inferiorly, and is attached above to the os unguis, and to the ridge on the superior maxillary and palate bones. It perfects the nasal duct inferiorly.

OS UNGUIS.

Situation, inner and anterior part of orbit.

Edges, serrated to join the os frontis above, the maxillary bone below, the nasal bone anteriorly, and the ethmoid bone behind.

External surface, divided by a perpendicular ridge, presents a groove anteriorly for the lachrymal sac, and a smooth surface behind to assist in forming the orbit.

Internal surface, covers the anterior ethmoidal cells. It is sometimes, although rarely, wanting.

NASAL BONES.

Situation, beneath the nasal process of frontal bone and between the nasal processes of superior maxillary bones.

External surface, convex, presents small foramina for blood-vessels.

Internal surface, concave, and grooved by the nasal nerves.

Superior edge, thick and serrated, to join the frontal bone.

Inferior edge, thin and expanded, joins the lateral nasal cartilages.

External edge is the longest, and is serrated to join the superior maxillary bone.

Internal edge is flat, and joins its fellow.

The nasal bone articulates with the frontal superior maxilla, opposite bone, ethmoid bone, and nasal cartilage.

VOMER.

Situation, in the median line of nose.

Superior edge, grooved to receive the azygos process of the sphenoid bone.

Anterior edge or lamella, slightly grooved to receive the nasal plate and nasal cartilage.

Posterior edge, free, looks towards the pharynx.

Inferior edge, the longest, is received into the nasal crest of the superior maxillary and palate bones.

INFERIOR MAXILLARY BONE.

Body, the anterior portion, projects inferiorly into *mental process or chin*, superiorly is surmounted by alveoli of four incisor teeth,—anteriorly has on each side a depression for muscles, and posteriorly, two depressions for the digastric muscles, above which are two pair of spines; the inferior for the geniohyoidei, and the superior for the genio-hyo-glossi muscles.

Symphysis, a vertical ridge in the centre of body.

Horizontal rami. On the outer surface of each is an oblique ridge for muscles; on the inner is the mylo-hyoidean ridge, above which is a depression for the sublingual gland, and another below for the sub-maxillary gland. The lower edge or base is rounded and grooved for the facial artery, and upon the upper edge are alveolar processes.

Angle, obtuse and rough for muscles.

Ascending rami, are thick and round posteriorly—externally smooth—internally grooved.

Coronoid process, passes upwards from anterior part of ascending ramus.

Condyle, transversely oblong to articulate with temporal bone.

Neck, constricted part below condyle.

Sigmoid notch, between condyle and coronoid process.

Inferior dental foramen, situated at internal surface of ascending ramus, and surmounted by a *spine*.

Mental foramen, situated at anterior surface of bone external to body.

Dental canal, traverses bone between the two foramina, and communicates with each alveolus. This canal contains the inferior dental nerve and artery.

OS HYOIDES.

Situation, in the anterior part of the neck between the chin and larynx.

Body, square, is rough anteriorly for muscles, smooth and concave posteriorly.

Greater cornua, pass obliquely backwards from the sides of the body, and end in tubercles.

Lesser cornua, are very small, and pass obliquely backwards and upwards from the point of junction of the great cornua and body, and connect the bone with the styloid process of temporal bone by means of the stylo-hyoid ligaments.

The bones of the ear are described in connection with the organ of hearing.

SUTURES OF THE CRANIUM AND FACE.

Coronal suture, commences a little behind the external angular process of the frontal bone, at the upper termination of the great wing of the sphenoid bone, and, inclining backwards, extends across the cranium to the opposite corresponding point, connecting in its course the frontal to the parietal bones.

Lambdoid suture, commences at the union of

the petrous portion of the temporal bone with the parietal and occipital bones, and extending across the posterior part of the cranium to the opposite corresponding point, connects the occipital to the parietal bones.

Sagittal suture, extends from the angle of the occipital bone forwards, connecting in its course the parietal bones and corresponding to the median line; it generally terminates in the coronal suture, but is occasionally prolonged to the nasal bones, dividing the frontal bone.

Squamous suture, corresponds to the semicircular edge of the squamous portion of the temporal bone, and connects it to the great wing of the sphenoid bone, and to the parietal bone.

Additamentum suturæ lambdoidalis, extends from the termination of the lambdoid suture to the foramen lacerum posterius, and unites the mastoid process of the temporal bone to the occipital.

Additamentum suturæ squamosæ, extends nearly horizontally backwards from the posterior termination of the squamous suture to the lambdoid suture, connecting the upper extremity of the mastoid portion of the temporal bone to the parietal bone.

Sphenoid suture, extends around the irregular margins of the sphenoid bone connecting it to all the bones of the head, and to the malar, superior maxillary, and palate bones.

Ethmoid suture, surrounds the ethmoid bone, connecting it to the frontal, nasal, superior maxillary, lachrymal, and palatine bones, and to the vomer.

Transverse suture, connects the bones of the face to those of the head.

Zygomatic suture, corresponds to the junction of the temporal with the malar bone.

ORBITS.

The orbits are two pyramidal cavities, the bases of which look outwards and forwards, and their apices in the contrary direction; so that two lines passing through their axes if prolonged posteriorly would decussate at the sella turcica. Each orbit is formed of seven bones, three of which—viz., the frontal, sphenoid, and ethmoid, are common to both orbits; the other four—viz., the lachrymal, superior maxillary, malar and palate bones, belonging to the orbit of their corresponding side. The roof is formed by the lesser wing of the sphenoid bone and the orbital plate of the frontal. At the outer angle is a depression for the lachrymal gland; at the inner a depression for the pulley of the trochlearis muscle. The floor is formed by the superior maxilla, the malar and palate bones, and contains the infra-orbital canal. The outer wall is formed by the malar and sphenoid bones; it is pierced by some small foramina for nerves; the inner is formed by the os planum of the ethmoid, the os unguis, superior maxilla, and a bit of the sphenoid. It is perforated by the anterior and posterior ethmoidal foramina. The foramina in the base of the orbit are, the supra-orbital, the infra-orbital, and the upper orifice of the nasal duct; within the orbit are the optic foramen, the superior foramen lacerum, and the spheno-maxillary fissure. Through the optic foramen passes the optic nerve and ophthalmic artery. Through the fissura lacera the third, fourth, ophthalmic division of the fifth nerve, the sixth nerve, and the ophthalmic vein.

TEMPORAL FOSSA,

Placed on the side of the cranium, is bounded by the frontal, sphenoid, malar, parietal, and temporal bones, and lodges the temporal muscle.

ZYGOMATIC FOSSA,

Extends from the temporal fossa downwards, and is bounded by the zygomatic arch, the superior maxillary bone, and the portion of the great wing of the sphenoid bone below its crest. Between the great wing of the sphenoid and the border of the superior maxillary bone is the spheno-maxillary fissure, which opens into the orbit.

PTERYGO-MAXILLARY FISSURE,

Lies deep in the zygomatic fossa, is bounded by the pterygoid processes, the tuberosity of the superior maxilla, and the nasal plate of the palate bone; and communicates with the spheno-maxillary fissure. The angle of union of the spheno-maxillary and pterygo-maxillary fissures is called the pterygo-maxillary fossa, into which five foramina open; foramen rotundum, for second branch of fifth cerebral nerve; foramen pterygoideum for the vidian nerve; pterygo-palatine for a small artery; posterior palatine for nerves and arteries; spheno-palatine, which lodges the spheno-palatine ganglion.

ARTICULATIONS.

TEMPORO-MAXILLARY ARTICULATION

Bony formation. Glenoid cavity of temporal bone and condyle of inferior maxillary bone.

External lateral ligament,—origin, zygomatic process and external meatus. *Insertion,* outside of the neck of condyle of lower jaw.

Internal lateral ligament,—origin, spinous process of sphenoid bone. *Insertion,* spinal orifice of inferior dental canal.

Stylo-maxillary ligament,—origin, styloid process of temporal bone. *Insertion,* angle of inferior maxilla.

Synovial Membranes; two; one is reflected from the cartilaginous surface of zygomatic eminence and glenoid cavity over the superior surface of the fibro-cartilage; the other covers the under surface of the fibro-cartilage, and is reflected over the condyle.

Inter-articular, or *fibro-cartilage*, of oval figure, thick in its circumference, thin in the centre, where it is sometimes perforated. Divides the joint into two.

Capsular ligament,—*origin*, zygomatic eminence, and glenoid fissure. *Insertion*, neck of lower jaw.

OCCIPITO-ATLANTOID ARTICULATION.

Bony formation, condyles of occipital bone and superior articular process of atlas.

Capsular ligament, imperfect.

Synovial membranes, cover the opposed cartilaginous surfaces.

Anterior occipito-atloidean ligament,—*origin*, anterior edge of foramen magnum. *Insertion*, upper edge of atlas, anterior to its articular processes.

Posterior occipito-atloidean ligament,—*origin*, posterior edge of foramen magnum. *Insertion*, upper edge of atlas behind its articular processes.

OCCIPITO-AXOID ARTICULATION.

Check, or *oblique ligaments*,—(*syn.*) *Ligam. alaria*; *odontoid ligaments*,—*origin*, from each side of odontoid process. *Insertion*, inner side of each occipital condyle.

Apparatus ligamentosus, or *occipito-axoidean*

ligament,—origin, lower part of basilar process, being posterior to odontoid process. *Insertion*, superior part of transverse ligament of atlas and bodies of second and third vertebræ, where it is continuous with the posterior common ligament.

ATLANTO-AXOID ARTICULATION.

Bony formation.—The anterior portion of the spinal hole of the atlas, and the odontoid process of axis; no intervertebral substance.

Anterior and posterior ligaments, as in all the other vertebræ.

Transverse ligament, attached on each side to inner edge of articular process of the atlas, and by means of apparatus ligamentosus to basilar process above, and body of axis below.

Synovial membranes.—One between posterior surface of odontoid process, and anterior surface of transverse ligament. Another covers the opposed cartilaginous surfaces of the atlas and odontoid process.

COMMON VERTEBRAL ARTICULATION.

Bony formation.—Opposed surfaces of the bodies and articulating process of the vertebræ.

Anterior vertebral ligament, extends from axis to sacrum, adhering to the bones and the inter-vertebral substances.

Posterior vertebral ligament, extends along the posterior part of bodies of vertebræ within the spinal canal. It is separated from the bodies of the vertebræ by some large veins.

Inter-vertebral ligaments, or fibro-cartilages, are placed between the bodies of all the vertebræ, except the atlas and dentata, and united to their flat surface above and below. They are thicker in front than behind in the neck and loins, and the contrary in the back: in the dorsal region they are attached to the heads of the ribs.

Synovial membranes and ligamentous fibres connect the articulating processes.

Ligamenta sub-flava, composed of yellow elastic tissue, are situated between the laminæ of the vertebræ from the second to the sacrum, completing the posterior part of the spinal canal.

Supra-spinous and inter-spinous ligaments, connect the spinous processes of the vertebræ.

Inter-transverse ligaments, connect the transverse processes. The Ligamentum Nuchæ extends from 7th cervical vertebra to the occiput.

COSTO-SPINAL ARTICULATIONS.

Bony formation.—Heads, and tubercles of ribs, and bodies and transverse processes of vertebræ.

Anterior ligament,—origin, front of head of rib. *Insertion*, side of the vertebra above and below, and to the inter-vertebral substance by radiating fibres, hence sometimes called "stellate ligament."

Inter-articular ligaments,—origin, projecting ridge in the articular surface of each rib. *Insertion*, cavity in the inter-vertebral substance in which the head is received. The upper and lower divisions of this joint have distinct synovial membranes.

Costo-transverse ligaments, three in number.

Internal costo-tranverse ligaments — origin, neck of each rib. *Insertion*, transverse process of the vertebra above.

Posterior costo-transverse ligaments, connect the tubercle of each rib to the corresponding transverse process.

Middle or interosseous costo-transverse ligament connects the neck of the rib to the contiguous transverse process: to see it, the bones must be forcibly separated.

Synovial membranes are between the tubercles and transverse processes.

CHONDRO-STERNAL ARTICULATION.

The ribs are connected to the sternum through the intervention of their cartilages, which form a joint lined by synovial membrane and secured anteriorly and posteriorly by ligamentous fibres.

LUMBO-SACRAL ARTICULATION.

Bony formation.—Last lumbar vertebra and

sacrum. These are joined together in the same manner as the other vertebræ.

Ilio-lumbar ligament,—origin, transverse processes of fifth lumbar vertebra. *Insertion,* posterior superior spinous process and crest of ilium.

Sacro-vertebral ligament extends almost vertically downwards from transverse process of last lumbar vertebra to base of sacrum.

ILIO-SACRAL ARTICULATION.

The ilium and sacrum are connected anteriorly and posteriorly by short ligamentous fibres. *Great sacro-sciatic ligament,—origin,* posterior inferior spine of ilium and side of sacrum and coccyx. *Insertion,* lower edge of tuber ischii. Along the ramus of the ischium it sends a falciform process which covers the obturator internus muscle.

Lessor sacro-sciatic ligament,—origin, side of sacrum and coccyx. *Insertion,* spine of ischium.

These ligaments convert the great lateral pelvic notch into foramina, called the greater and lesser sacro-sciatic foramina. The former, superior, transmits the pyramidalis muscle, the great sciatic nerve, the sciatic vessels, and the gluteal vessels and nerve; the latter, inferior, and bounded by the two ligaments, transmits the tendon of the obturator internus muscle and the internal pudic vessels and nerve.

SACRO-COCCYGEAL ARTICULATION.

The sacrum and coccyx are united together by a similar substance to the inter-vertebral, and by ligamentous bands anteriorly and posteriorly.

PUBIC ARTICULATION.

Fibro-cartilage, which attaches closely the bones of the pubes; also ligamentous fibres.

Sub-pubic ligament, passes from the ramus of one bone to the other, and rounds off the angle formed by their union.

Obturator ligament, attached to the circumference of obturator foramen, except superiorly, where the obturator nerve and vessels pass.

STERNO-CLAVICULAR ARTICULATION.

Anterior ligament,—*origin*, anterior surface of sternal end of clavicle. *Insertion*, anterior surface of sternum.

Posterior ligament,—*origin*, posterior surface of sternal end of clavicle. *Insertion*, back part of sternum.

Costo-clavicular ligament,—*origin*, lower surface of sternal end of clavicle. *Insertion*, cartilage of first rib.

Inter-clavicular ligament extends from the sternal extremity of one clavicle to that of the other, dipping down in its course to become attached to the upper surface of the sternum.

Inter-articular cartilage, thin below and attached to sternum, thick above and attached to clavicle; having a synovial membrane connected to each surface and its corresponding bone.

SCAPULO-CLAVICULAR ARTICULATION.

Superior acromio-clavicular ligament,—origin, upper surface of acromion. *Insertion,* upper part of clavicle.

Inferior acromio-clavicular ligament.—Attached to under surface of each bone.

Synovial membrane covers the articulating surfaces in the usual manner.

CORACO-CLAVICULAR ARTICULATION.

Conoid ligament, triangular; base connected to the tubercle on inferior surface of clavicle, apex at the broad part of coracoid process.

Trapezoid ligament, attached above to an oblique line on the clavicle; below to upper part of the coracoid process.

LIGAMENTS OF THE SCAPULA.

Coraco-acromial ligament,—origin, broad from coracoid process. *Insertion,* narrow into point of acromion.

Posterior or coracoid ligament,—origin, superior costa of scapula behind the notch. *Insertion,* base of coracoid process. This ligament converts the notch into a foramen, through which passes the suprascapular nerve.

HUMERO-SCAPULAR ARTICULATION.

Bony formation.—Head of humerus and glenoid cavity of scapula.

Capsular ligament,—origin, circumference of neck of scapula. *Insertion,* around the neck of humerus. It is strengthened by the coraco-humeral ligament, and by prolongations from the tendons of the supra and infra spinatus and teres minor muscles.

Coraco-humeral ligament,—origin, coracoid process. *Insertion,* anterior part of great tuberosity.

Synovial membrane is reflected over the surface of the glenoid cavity around the glenoid ligament; lines the capsular ligament, head of humerus, and bicipital groove.

Glenoid ligament, closely connected with the tendon of the biceps flexor cubiti, encircles the glenoid cavity, and by elevating the border, adds to the depth of the articulating surface.

HUMERO-CUBITAL ARTICULATION.

Bony formation, articular processes of humerus, great sigmoid cavity of ulna, head of radius.

External lateral ligament,— origin, external condyle of humerus. *Insertion,* annular ligament of radius.

Internal lateral ligament,—origin, internal condyle. *Insertion,* inner edge of olecranon and coronoid processes.

Anterior ligament consists of thin fibres,— *origin*, principally from the above internal condyle and depression on the forepart of humerus; *insertion*, annular ligament of radius and synovial membrane.

Posterior ligament is composed of fibres which extend from one condyle to the other, and are attached to the synovial membrane.

Synovial membrane is reflected from behind the anterior ligament to neck of radius and annular ligament; it then lines the sigmoid cavities of the ulna, and is reflected to the lateral ligaments and tendon of the triceps muscle, which conducts it to the posterior depression of the humerus; it is then expanded over its articular processes.

SUPERIOR RADIO-ULNAR ARTICULATION.

Bony formation, lesser sigmoid cavity of ulna and inner side of head of radius.

Coronary ligament—origin, anterior border of lesser sigmoid cavity of ulna; *insertion*, posterior border of the same cavity. It encircles the neck of radius.

Oblique ligament—origin, coronoid process of ulna. *Insertion*, radius below its tubercle.

Interosseous ligament connects the opposed edges of radius and ulna, its fibres descending obliquely inwards from the former bone to the latter.

INFERIOR RADIO-ULNAR ARTICULATION.

Bony formation, round head of ulna, and sigmoid cavity of radius; the bones are united by some fibrous bands, which are so thin as not to need description.

Fibro-cartilage—origin, styloid process of ulna. *Insertion*, inner edge of radius below the ulna.

The synovial membrane, called sometimes membrana sacciformis, extends horizontally between the extremity of the ulna and the fibro-cartilage, and vertically between the opposed articulating surfaces of the radius and ulna.

RADIO-CARPAL ARTICULATION.

Bony formation, lower end of radius; scaphoid, lunar, and cuneiform bones.

The extremity of the ulna is separated from the joint by the fibro-cartilage.

External lateral ligament—origin, styloid process of radius. *Insertion*, scaphoid bone, and by some fibres into annular ligament and trapezium.

Internal lateral ligament—origin, styloid process of ulna. *Insertion*, cuneiform bone and pisiform bone.

Posterior ligament—origin, posterior part of radius and fibro-cartilage. *Insertion*, back part of superior row of carpus.

Anterior ligament—origin, anterior part of

radius and fibro-cartilage. *Insertion*, fore part of first row of carpus.

The synovial membrane lines the head of the radius, the fibro-cartilage, and is then reflected over the three carpal bones.

CARPAL ARTICULATIONS.

The bones of the carpus are articulated by ligamentous bands, both anteriorly and posteriorly.

The scaphoid, lunar, and cuneiform bones are connected by dorsal and palmar ligaments, extending from bone to bone: between the opposed surfaces are interosseous fibro-cartilages.

The pisiform bone possesses a distinct fibrous capsule and synovial membrane.

The trapezium, trapezoides, magnum, and unciform, are also connected by dorsal and palmar ligaments.

Interosseous fibro-cartilages are found on either side of the os magnum.

Lateral ligaments extend on the radial side from the scaphoid to the trapezium; on the ulnar side, from the cuneiform to the unciform.

The carpal synovial membrane extends between the first and second row of carpal bones, and is prolonged to the carpal extremities of the four inner metacarpal bones. The trapezium articulates with the metacarpal bones of the thumb by a distinct capsule and synovial membrane.

Annular ligament—origin, trapezium and scaphoid bones. *Insertion,* cuneiform and unciform and pisiform bones.

CARPO-METACARPAL ARTICULATION.

The carpus and metacarpus are secured by fibrous bands, which pass in different directions, and cover the synovial membranes.

METACARPO-PHALANGEAL ARTICULATIONS.

The heads of the metacarpus and first phalanges are secured by *lateral ligaments*, and are lined by *synovial membranes;* a transverse ligament connects the digital extremities of the metacarpal bones one with another.

INTER-PHALANGEAL ARTICULATIONS.

The phalanges are connected to each other by means of two lateral ligaments, one anterior ligament, and between each of their joints is a synovial membrane.

ILIO-FEMORAL ARTICULATION.

Bony formation, acetabulum and head and part of neck of femur.

Cotyloid ligament, a fibro-cartilaginous circular band adhering to the margin of the acetabulum.

Transverse ligament, attached to the opposite

points of the notch of the acetabulum, and partly filling it up.

Capsular ligament—*origin*, circumference of acetabulum and transverse ligament. *Insertion*, below root of trochanter major, and the two inter-trochanteric lines.

Accessory or ilio-femoral ligament—*origin*, anterior inferior spinous process of ilium. *Insertion*, anterior trochanteric line.

Synovial membrane, reflected from inside of capsule upon periosteum of neck, and cartilaginous surface of head; is continued over inter-articular ligament, and thence is reflected upon the cartilaginous surface of the acetabulum.

Inter-articular ligament, or *ligamentum teres*—*origin*, depression on the head of femur. *Insertion*, by two bands into the extremities of the notch, and into the transverse ligament.

FEMORO-TIBIAL ARTICULATION.

Bony formation, condyles of femur, head of tibia and the patella.

Ligamentum patellæ—*origin*, lower angle of patella. *Insertion*, tubercle of tibia.

Posterior ligament—*origin*, tendon of semi-membranosus muscle at internal and posterior part of tibia. *Insertion*, external condyle of femur.

Internal lateral ligament—*origin*, internal condyle of femur. *Insertion*, internal condyle of tibia and semilunar cartilage.

External lateral ligament—origin, external condyle of femur. *Insertion*, head of fibula. This ligament is often divided into two by the tendon of the biceps muscle.

Synovial membrane lines the back part of the patella, from which it is reflected two or three inches on the fore part of the femur, and on its condyles; from thence it is conducted by the crucial ligaments to the semilunar cartilages, and head of tibia.

Alar ligaments arise from each side of patella, and unite below the bone. They are mere folds of synovial membrane.

Ligamentum mucosum—origin, fatty substance behind ligamentum patellæ. *Insertion*, hollow between the condyles. It also is a fold of synovial membrane.

Transverse ligament, attached to the anterior portion of each semilunar cartilage.

Anterior crucial ligament—origin, inner side of external condyle. *Insertion*, near the fore part of head of tibia, where it is connected to the cornu of the internal semilunar cartilage.

Posterior crucial ligament—origin, outer side of internal condyle. *Insertion*, depression on back part of head of tibia, and external semilunar cartilage.

Semilunar cartilages, thick externally, thin internally: concave above, flat below. The outer convex edge of the internal is attached to the lateral ligament; the inner edge is free; the anterior and posterior extremities of each are attached to the head of the tibia. The outer cartilage is circular, the inner is semi-circular.

TIBIO-FIBULAR ARTICULATION.

The head of the fibula is attached to the tibia by *anterior* and *posterior fibrous bands* and synovial membrane.

Interosseous membrane extends from one bone to the other, nearly the whole length.

The lower extremities of the tibia and fibula are connected together by *anterior* and *posterior* ligaments; the synovial membrane is a prolongation of that which lines the ankle joint.

ARTICULATION OF THE ANKLE.

Bony formation, lower ends of tibia, fibula, and astragalus.

Internal lateral ligament—origin, internal malleolus. *Insertion*, astragalus, naviculare, and os calcis.

External lateral ligament has three parts; all take their *origin* from the external malleolus. *Insertion* of *anterior*, upper and outer part of astragalus. *Insertion* of *middle*, os calcis. *Insertion* of *posterior*, ridge on the back of astragalus, which bounds the groove for the flexor longus pollicis.

The synovial membrane covers the opposed surfaces of the bones, and sends upwards a prolongation into the inferior tibio-fibular articulation.

ARTICULATIONS OF THE TARSUS.

The astragalus and os calcis have two articular surfaces, covered by synovial membranes.

Interosseous ligament passes nearly perpendicularly from the groove which separates the inferior articular surfaces of the astragalus, to the corresponding groove in the os calcis.

Posterior ligament is attached to the posterior edges of the astragalus and os calcis.

External lateral ligament passes from the astragalus to the outer surface of the os calcis.

There are two synovial membranes: one posterior to the interosseous ligament, between the astragalus and os calcis; a second anterior to the interosseous ligament, between the astragalus and os calcis, and continued forwards between the astragalus and os naviculare, over the calcaneo-navicular ligament.

The bones of the tarsus are connected on their dorsal and plantar aspects by numerous ligamentous bands.

Calcaneo-scaphoid ligament—*origin*, inferior surface of os calcis. *Insertion*, inferior surface of os naviculare.

It is composed of elastic tissue, and is supported inferiorly by the tendon of the tibialis posticus muscle: superiorly it forms with the os calcis and os naviculare a cup, which receives the head of the astragalus.

Inferior calcaneo-cuboid ligament—*origin*, posterior inferior part of os calcis. *Insertion*, under part of cuboid bone, and third and fourth metatarsal bones.

The superficial fibres, longer than the deeper, form the "ligamentum longum plantæ." There is a distinct synovial membrane between the calcaneum and cuboid bones.

The three cuneiform bones are connected by dorsal, plantar, and interosseous ligaments: one synovial membrane is reflected over their opposed surfaces with the os naviculare.

TARSO-METATARSAL ARTICULATIONS.

These joints are secured by dorsal and plantar ligaments, and are lined by synovial membranes. The metatarsal bones are secured to the phalanges, and the phalanges to each other, by lateral ligaments and synovial membranes. There is one synovial membrane between the internal cuneiform bone and the first metatarsal bone: a second between the os cuboides and the fourth and fifth metatarsal bones. The second and third metatarsal bones form part of the articulation between the cuneiform and the navicular bones.

MUSCLES.

HEAD.

Occipito-frontalis. — *Origin,* two external thirds of superior semicircular ridge of occipital bone, and posterior external part of mastoid process of temporal bone. *Insertion*, integuments of eyebrows, when the fibres become blended with those of the pyramidalis nasi, the corrugator supercilii, and the orbicularis palpebrarum.

MUSCLES OF EXTERNAL EAR.

Attollens aurem.—*O.* cranial aponeurosis above external ear. *I.* upper and anterior part of cartilage of ear.

Attrahens aurem.—*O.* posterior part of zygomatic process and cranial aponeurosis. *I.* anterior part of helix.

Retrahens aurem.—*O.* mastoid process. *I.* back part of concha.

FACE.

Orbicularis palpebrarum.—*O.* Internal angular process of os frontis and upper edge of tendo palpebrarum. *I.* nasal process of superior maxillary bone and inferior edge of tendo palpebrarum.

Tensor tarsi.—*O.* posterior edge of os unguis. *I.* lachrymal ducts as far as puncta.

Corrugator supercilii.—*O.* internal angular process of os frontis. *I.* middle of eyebrow.

Pyramidalis nasi.—*O.* occipito-frontalis muscle, descends along nasal bones. *I.* compressor nasi muscle.

Compressor nasi.—*O.* canine fossa in superior maxilla. *I.* dorsum of nose.

Levator labii superioris alæque nasi.—*O.* upper extremity of nasal process of superior maxilla, and from edge of orbit above infra-orbital hole. *I.* ala nasi, upper lip, and orbicularis oris muscle.

Zygomaticus minor.—*O.* upper part of malar bone. *I.* upper lip, near commissure. Sometimes wanting.

Zygomaticus minor.—*O.* lower part of malar bone, near zygomatic suture. *I.* angle of mouth.

Levator anguli oris.—*O.* canine fossa above alveola of the first molar tooth. *I.* commissure of lips and orbicularis oris.

Depressor labii superioris alæque nasi.—*O.* myrtiform fossa above canine and incisor teeth of superior maxilla. *I.* integuments of upper lip and fibro-cartilage of septum and ala nasi.

Depressor anguli.—*O.* external oblique line on lower jaw, extending from anterior edge of masseter muscle to mental foramen. *I.* commissure of lips.

Depressor labii inferioris.—*O.* side and front of lower maxilla, above its base. *I.* half of lower lip and orbicularis oris.

Levator labii inferioris.—*O.* alveoli of incisor teeth of lower jaw. *I.* integument of chin.

Orbicularis oris surrounds mouth by two fleshy fasciculi.

Buccinator.—*O.* two last alveoli of superior maxilla and external surface of posterior alveoli of lower maxilla, and from a ligament (pterygomaxillary) which extends from the internal pterygoid plate of the sphenoid bone to the extremity of the mylo-hyoid ridge of the inferior maxilla. *I.* commissure of lips.

MUSCLES OF LOWER JAW.

Masseter consists of two portions: *Anterior portion.* *O.* superior maxilla where it joins malar bone, and inferior edge of latter. *I.* outer surface of angle of lower jaw.

Posterior portion.— *O.* edge of malar bone, and zygomatic arch, as far as glenoid cavity. *I.* external side of angle and ramus of lower jaw.

Temporal.— *O.* side of cranium, beneath semicircular ridge on parietal bone, temporal fossa, and aponeurosis. *I.* coronoid process of inferior maxilla to last molar tooth.

Pterygoideus internus.— *O.* internal pterygoid plate and pterygoid fossa of palate bone. *I.* inner side or angle of jaw and rough surface above.

Pterygoideus externus.— *O.* outer side of external pterygoid plate, spinous process of sphenoid bone and tuberosity of superior maxilla. *I.* anterior and internal part of neck of lower jaw, cartilage, and capsular ligament.

MUSCLES OF ANTERIOR AND LATERAL PARTS OF NECK.

Platysma-myoides.— *O.* cellular membrane covering upper and outer part of deltoid and great pectoral muscles. *I.* chin; fascia along side of lower jaw; and fascia covering parotid gland. Some of the fibres become blended with those of the depressor labii inferioris and anguli oris.

Sterno-cleido mastoideus.—*O.* upper and anterior part of the first bone of sternum and sternal third, sometimes half, of clavicle. *I.* upper part of mastoid process and external third of superior transverse ridge of occipital bone.

Sterno-hyoideus.—*O.* posterior surface of first bone of sternum, cartilage of first rib, sternal end of clavicle, and sterno-clavicular articulation. *I.* lower border of body of os hyoides.

Sterno-thyroideus.—*O.* posterior surface of first bone of sternum and cartilage of second rib. *I.* oblique line on ala of thyroid cartilage. It is broader than the preceding.

Omo hyoideus.—*O.* behind semilunar notch in scapula, from the ligament which passes over the notch, and from base of coracoid process. *I.* lower border of os hyoides, at the junction of its body and great cornu. It is a double-bellied muscle, the mesial tendon being bound down by a pulley of cervical fascia.

Digastricus.—A double-bellied muscle. Posterior belly. *O.* groove internal to mastoid process. Anterior belly. *O.* rough depression on inner surface of base of jaw, near its symphysis. The two bellies unite at an angle, in a tendon which passes through the fibres of the stylo-hyoid muscle, and is inserted into the os hyoides.

Mylo-hyoideus. *O.* mylo-hyoid ridge of inferior maxilla. *I.* base of os hyoides, chin, and middle tendinous line common to it and its fellow.

Genio-hyoideus. *O.* inner side of chin, above the digastricus. *I.* base of os hyoides.

Hyo-glossus. *O.* great cornu and part of body of os hyoides. *I.* side of tongue.

Genio-hyo-glossus. *O.* eminence inside chin, below frænum linguæ. *I.* mesial line of tongue from apex to base, and body and lesser cornu of os hyoides.

Lingualis consists of fasciculi of fibres, running from base to apex of tongue, and lying between the genio-hyo-glossus, and the hyo and stylo glossi.

Stylo-hyoideus. *O.* outer side of styloid process, near its middle. *I.* cornu and body of os hyoides.

Stylo-glossus. *O.* styloid process, near its tip and the stylo-maxillary ligament. *I.* side of tongue, as far as the tip.

Stylo-pharyngeus. *O.* inner part of root of styloid process. *I.* side of pharynx, cornu of os hyoides, and thyroid cartilage.

MUSCLES OF THE PHARYNX.

Constrictor pharyngis inferior. *O.* side of cricoid cartilage, inferior cornu, and posterior edge of thyroid cartilage. *I.* with its fellow, along mesial line on back of pharynx.

Constrictor pharyngis medius. *O.* great cornu

and also lesser cornu of os hyoides, stylo-hyoid and thyro-hyoid ligaments. *I.* mesial tendinous line, and basilar process of occipital bone.

Constrictor pharyngis superior. *O.* petrous portion of temporal bone, lower part of internal pterygoid plate, also pterygo-maxillary ligament, posterior third of mylo-hyoid ridge, and side of base of tongue. *I.* basilar process of occipital bone and mesial line of pharynx.

MUSCLES OF THE PALATE.

Levator palati. *O.* petrous portion of temporal bone in front of foramen caroticum, and from cartilage of Eustachian tube. *I.* broad, into the velum.

Tensor, vel circumflexus palati. *O.* spinous process of sphenoid, and fore part of Eustachian tube; tendon turns round hamular process. *I.* into velum, meeting its fellow in the mesial line.

Levatores uvulæ. *O.* posterior extremity of spine of palate bones. *I.* cellular tissue of uvula.

Palato-glossus. *O.* inferior surface of velum. *I.* side of tongue. It forms anterior half arch.

Palato-pharyngeus. *O.* inferior surface of palate. *I.* side and back of pharynx, and superior cornu of thyroid cartilage. It forms the posterior half arch.

LARYNX.

Thyro-hyoideus. *O.* oblique ridge on ala of thyroid cartilage. *I.* lower edge of great cornu of os hyoides.

Crico-thyroideus. O. fore part of cricoid cartilage. I. lower border of thyroid cartilage.

Thyro-arytenoideus. O. posterior surface of thyroid cartilage, near its angle. I. anterior edge of arytenoid cartilage.

Crico-arytenoideus lateralis. O. upper edge of side of cricoid cartilage. I. base of arytenoid cartilage.

Crico-arytenoideus posticus. O. depression on posterior surface of cricoid cartilage. I. outer side of base of arytenoid cartilage.

Arytenoideus fills the interval between arytenoid cartilages, and consists of two arrangements of fibres: *oblique*, run from apex of one cartilage to base of opposite one; *transverse*, are attached to posterior surface of each cartilage.

DEEP MUSCLES ON ANTERIOR AND LATERAL PARTS OF THE NECK.

Longus colli. Divided into two portions, vertical and oblique. Vertical portion. O. bodies of three upper dorsal and four lower cervical vertebræ, and the intervertebral fibro-cartilages. I. bodies of second and third cervical vertebræ. Oblique portion. O. anterior part of transverse processes of third, fourth, and fifth cervical vertebræ. I. body of the atlas.

Rectus capitis anticus major. O. anterior tubercles of transverse processes of four inferior cervical vertebræ. I. basilar process of occipital bone.

Rectus capitis anticus minor. *O.* transverse process of atlas. *I.* basilar process of occipital bone.

Rectus capitis lateralis. *O.* transverse process of atlas. *I.* jugular process of occipital bone.

Scalenus anticus. *O.* anterior tubercles of transverse processes of third, fourth, fifth, and sixth cervical vertebræ. *I.* upper surface of first rib, just anteriorly to its middle.

Scalenus medius. *O.* posterior tubercles of transverse processes of four or five inferior cervical vertebræ. *I.* upper edge of first rib.

Scalenus posticus. *O.* posterior tubercles of two or three inferior cervical vertebræ. *I.* upper edge of second rib between its tubercle and angle.

THORAX.

Pectoralis major. *O.* sternal half of clavicle, anterior surface of sternum, cartilages of second, third, fourth, fifth, and sixth ribs, and aponeurosis common to it and external oblique muscle. *I.* by a flat tendon into anterior edge of bicipital groove, and by an aponeurosis into fascia of forearm.

Pectoralis minor. *O.* external surfaces and upper edges of third, fourth, and fith ribs, sometimes from second. *I.* inner and upper surface of coracoid process of scapula.

Subclavius. *O.* cartilage of first rib. *I.* external half of inferior surface of clavicle.

Serratus magnus. *O.* by eight or nine fleshy

slips, from eight or nine superior ribs. *I.* base of scapula.

Intercostales are twenty-two in number on each side: eleven *internal* and eleven *external*. The fibres of the external pass obliquely from behind, forwards and downwards; those of the internal in the opposite direction.

External. *O.* inferior edge of each rib, commencing at transverse processes of vertebræ. *I.* external lip of superior edge of rib beneath, extending to costal extremities of cartilages.

Internal. *O.* at sternum from the inner lip of lower edge of each cartilage and rib as far as angle. *I.* inner lip of superior edge of cartilage and rib beneath.

Levatores costarum. *O.* extremities of each dorsal transverse process. *I.* upper edge of rib below, between tubercle and angle.

Triangularis sterni. *O.* posterior surface and edge of lower part of sternum and ensiform cartilage. *I.* cartilages of fourth, fifth, and sixth ribs.

MUSCLES OF THE BACK.

First Layer.

Trapezius. *O.* internal third of superior transverse ridge of occipital bone, ligamentum nuchæ, and spinous process of last cervical and all dorsal vertebræ. *I.* posterior border of external third of clavicle, acromion process, and superior edge of spine of scapula.

Latissimus dorsi. *O.* Six inferior dorsal spines, and by lumbar fascia from all lumbar spines, from back of sacrum, posterior third of crest of ilium, and from three to four inferior ribs. *I.* posterior edge of bicipital groove of humerus.

Second Layer.

Rhomboideus minor. *O.* lower part of ligamentum nuchæ and last cervical spinous process. *I.* base of scapula, opposite to its spine.

Rhomboideus major. *O.* four or five superior dorsal spines. *I.* base of scapula from spine to inferior angle.

Levator anguli scapulæ. *O.* posterior tubercles of transverse processes of four or five superior cervical vertebræ. *I.* base of scapula, between spine and superior angle.

Serratus posticus superior. *O.* ligamentum nuchæ, and two or three dorsal spines. *I.* second, third, and fourth ribs, external to angles.

Serratus posticus inferior. *O.* two last dorsal and two superior lumbar spines. *I.* lower edges of four inferior ribs anterior to angles.

Splenius muscle, flat and oblique, single at its origin; divides at its insertion into two portions, the splenius colli, and splenius capitis; the former attached to the cervical vertebræ, the latter to the cranium. *O.* from spines of five superior dorsal and last cervical vertebra, and from ligamentum nuchæ as high as the third cervical vertebra. *I.* the lower portion, splenius colli, the smaller, into posterior tubercles of transverse process of three or four superior cervical vertebræ. Upper portion, splenius capitis, into posterior part of mastoid process, superior curved ridge of occipital bone, and the rough surface below it.

Third Layer.

The following muscles lie beneath the serrati, and a fascia called the vertebral aponeurosis. The fleshy mass occupying the vertebra grooves of either side is called erector spinæ, and it divides opposite the last rib into two portions, sacro-lumbalis, and longissimus dorsi.

Erector spinæ. *O.* from a dense fascia connecting with the spines of the sacrum, from the posterior third of the crista of the ilium, from the posterior surface of the sacrum, and from the sacro-sciatic ligaments. The outer portion, called sacro-lumbalis, is inserted into the angles of six or seven lower ribs. Musculus accessorius ad sacro-lumbalem, a continuation of sacro-lumbalis. *O.* angles of six or seven lower ribs, internal to the tendons of preceding muscle. *I.* angles of six upper ribs, and into transverse process of last cervical vertebra.

Cervicalis ascendens, the continuation of sacro-lumbalis into the neck. *O.* from angles of third, fourth, fifth, and sixth ribs. *I.* transverse processes of third, fourth, fifth, and sixth cervical vertebræ.

The inner portion of erector spinæ, longissimus dorsi, is inserted into transverse processes of all the dorsal and lumbar vertebræ and into all the ribs between their tubercles and angles. Transversalis colli, a continuation of the longissimus dorsi. *O.* transverse processes of third, fourth, fifth, and sixth dorsal vertebræ. *I.* posterior tubercle of transverse processes of second, third, fourth, fifth, and sixth cervical vertebræ.

Trachelo-mastoid, the prolongation of the longissimus dorsi to the head. *O.* transverse processes of four upper dorsal and four lower cervical vertebræ. *I.* posterior margin of mastoid process.

Complexus. *O.* transverse and oblique processes of three or four inferior cervical and five or six superior dorsal vertebræ. *I.* into occipital bone between its two transverse ridges.

Fourth Layer.

Semi-spinalis colli. *O.* extremities of transverse processes of five or six superior dorsal vertebræ. *I.* by four heads into spines of second, third, fourth, and fifth cervical vertebræ.

Semi-spinalis dorsi. *O.* by five or six tendons, from transverse processes of dorsal vertebræ, from fifth to eleventh. *I.* extremity of spines of two inferior cervical and three or four superior dorsal vertebræ.

Multifidus spinæ. *O.* first fasciculus arises from spine of vertebra dentata, and is inserted into transverse process of third, each successively in a similar manner down to the last, which arises from the spine of last lumbar vertebra, and is inserted into transverse process of sacrum.

Interspinales—are situated between spinous processes of vertebræ: they are in pairs in the cervical region.

Intertransversales—attached and situated as their name implies.

Rectus capitis posticus major. O. spinous process of second vertebra. I. inferior transverse ridge of os occipitis.

Rectus capitis posticus minor. O. posterior part of atlas. I. occipital bone, behind foramen magnum.

Obliquus capitis inferior. O. spinous process of second vertebra. I. extremity of transverse process of atlas.

Obliquus capitis superior. O. upper part of transverse process of atlas. I. occipital bone, between its transverse ridges, posterior to mastoid process.

UPPER EXTREMITY.

Shoulder Arm.

Deltoideus. O. lower edge of spine of scapula, anterior edge of acromion, and external third of clavicle. I. rough surface on outer side of humerus, near its centre.

Supra-spinatus. O. all scapula above the spine, which forms supra-spinous fossa, and from fascia covering muscle. I. upper and fore part of great tuberosity of humerus.

Infra-spinatus. O. inferior surface of spine and dorsum of scapula beneath, except near the neck, as low down as posterior ridge on inferior costa. I. middle of great tuberosity of humerus.

Teres minor. O. depression between the two ridges on inferior costa of scapula, from fascia covering it, and ligamentous septa. I. inferior depression on great tuberosity of humerus.

Subscapularis. O. all the surface and circumference of subscapular fossa. I. lesser tubercle of humerus, and a small portion of the neck of the bone.

Teres major. O. rough surface on inferior angle of scapula, below infra-spinatus. I. posterior edge of bicipital groove.

Coraco-brachialis. O. coracoid process and tendon of short head of biceps. I. internal side of humerus about its middle.

Biceps. O. short head, from coracoid process; long head, from upper edge of glenoid cavity. I. into tubercle of radius. It gives off an aponeurosis at the bend of the arm.

Brachialis anticus. O. centre of humerus by two slips on either side of insertion of deltoid, and fore part of humerus to its condyles. I. coronoid process of ulna and rough surface beneath.

Triceps extensor cubiti. O. long head, from lower part of neck of scapula and inferior costa. Second head, from ridge on humerus, below insertion of teres minor. Third head, from ridge below insertion of teres major, leading to the internal condyle, and from internal intermuscular ligament. *I.* olecranon process of ulna, and fascia of forearm.

Forearm and Hand.

Palmaris brevis, lies between the skin and the palmar fascia. *O.* annular ligament and palmar fascia. *I.* integuments on inner side of palm.

Pronator radii teres. *O.* anterior part of internal condyle, fascia of forearm, intermuscular septa, and by a small slip, separated from the larger head by the median nerve, from coronoid process of ulna. *I.* outer and back part of radius, about its centre.

Flexor carpi radialis. *O.* inner condyle and intermuscular septa. *I.* base of metacarpal bone of index finger.

Palmaris longus. *O.* inner condyle and fascia of forearm. *I.* annular ligament and palmar aponeurosis, near root of thumb.

Flexor carpi ulnaris. *O.* inner condyle, inner side of olecranon, a tendinous band between these points, under which passes the ulnar nerve, inner edge of nearly whole length of ulna, and fascia of forearm. *I.* os pisiform and base of fifth metacarpal bone.

Flexor digitorum sublimis perforatus. *O.* inner condyle, internal lateral ligament, coronoid process, and radius below tubercle. *I.* anterior part of second phalanges of each finger.

Flexor digitorum profundus perforans. *O.* upper three-fourths of anterior surface of ulna by two heads, which embrace the insertion of the brachialis anticus, internal half of interosseous ligament. *I.* last phalanx of each finger.

Flexor pollicis longus. *O.* fore part of radius below the tubercle, from interosseous membrane to within two inches of carpus, and from coronoid process. *I.* last phalanx of thumb.

Pronator quadratus. *O.* inferior fifth of anterior surface of ulna. *I.* anterior part of inferior fourth of radius.

Supinator radii longus. *O.* external ridge of humerus to within two inches of outer condyle, and from intermuscular ligament. *I.* rough surface on the outside of radius, near its styloid process.

Extensor carpi radialis longior. *O.* ridge of humerus, between supinator longus and external condyle. *I.* back part of the carpal extremity of metacarpal bone of index finger.

Extensor carpi radialis brevior. *O.* inferior and posterior part of external condyle, and external lateral ligament. *I.* carpal extremity of third metacarpal bone.

Extensor digitorum communis. *O.* external condyle, fascia of forearm and its intermuscular septa. *I.* posterior aspect of all the phalanges of four fingers.

Extensor carpi ulnaris. *O.* external condyle, fascia, and septa, and middle third of ulna. *I.* carpal end of fifth metacarpal bone.

Anconeus. *O.* posterior and inferior part of external condyle and lateral ligament. *I.* external surface of olecranon, and superior fifth of posterior surface of ulna.

Extensor minimi digiti. *O.* in common with and between extensor digitorum communis and extensor carpi ulnaris. *I.* posterior part of phalanges of little finger.

Supinator radii brevis. *O.* external condyle, external, lateral, and coronary ligaments, and from a ridge on outer side of ulna, which commences below its lesser sigmoid cavity. *I.* upper third of external and anterior surface of radius, from above its tubercle to the insertion of pronator radii teres.

Extensor ossis metacarpi pollicis, lies immediately below the border of the supinator radii brevis. *O.* middle of posterior part of ulna, interosseous ligament, and posterior surface of radius. *I.* base of metacarpal bone of thumb.

Extensor primi internodii pollicis. *O.* interosseous ligament, radius, and occasionally a small portion of the ulna. *I.* base of first phalanx.

Extensor secundi internodii pollicis. *O.* posterior surface of ulna, above its centre, and from interosseous ligament. *I.* base of last phalanx.

Extensor indicis. *O.* middle of posterior surface of ulna, and interosseous ligament. *I.* second and third phalanges, uniting with the tendon of the common extensor.

Abductor pollicis. *O.* anterior surface of annular ligament and os scaphoides. *I.* outside of base of first phalanx, and by an expansion into both phalanges.

Opponens pollicis. *O.* annular ligament and

os trapezium. *I.* anterior extremity of metacarpal bone of thumb.

Flexor pollicis brevis. *O. external head,* from inside of annular ligament and trapezium and sheath of the flexor carpi radialis. *I.* external sesamoid bone and base of first phalanx of thumb. *Internal head.* *O.* from os trapezoides, os magnum, and base of metacarpal bone of middle finger. *I.* internal sesamoid bone and base of first phalanx.

Adductor pollicis. *O.* three-fourths of anterior surface of the third metacarpal bone. *I.* inner side of base of first phalanx of thumb.

Abductor indicis. *O.* metacarpal bone of fore finger and one-half of that of the thumb. *I.* outer side of base of first phalanx. The radial artery passes between its two heads.

Lumbricales. *O.* outer side of the tendons of flexor profundus, near the carpus, a little beyond annular ligament. *I.* middle of first phalanx into tendinous expansion covering the back of each finger.

Abductor minimi digiti. *O.* annular ligament and os pisiforme. *I.* ulnar side of first phalanx.

Flexor brevis minimi digiti. *O.* annular ligament and unciform bone. *I.* base of first phalanx of little finger.

Abductor minimi digiti. *O.* internal to last, and overlapped by it. *I.* all the metacarpal bone of little finger.

Interossei palmares, three in number. *O.*

sides of metacarpal bones. *I.* first phalanges and tendinous expansion covering the dorsum of each finger.

1st, arises from ulnar side of second metacarpal bone; 2d, arises from radial side of fourth metacarpal bone; 3d, arises from radial side of fifth metacarpal bone; each, joining with the tendons of common extensor, is inserted into the base of the first phalanx of the corresponding finger.

Interossei dorsales, four in number. *O.* opposed sides of two metacarpal bones. *I.* base of first phalanx of each finger and posterior tendinous expansion.

1st, has been already described under the name of abductor indicis.

2d, arises from second and third metacarpal bones, inserted into radial side of base of first phalanx of middle finger.

3d, arises from third and fourth metacarpal bones, inserted into ulnar side of base of first phalanx of middle finger.

4th, arises from fourth and fifth metacarpal bones, inserted into ulnar side of base of first phalanx of ring finger.

All the palmar interossei are adductors, and all the dorsal interossei are abductors to a line drawn longitudinally through the middle finger.

ABDOMEN.

Obliquus externus vel descendens. *O.* external surfaces of eight or nine inferior ribs at a little distance from their cartilages. *I.* the fibres end in a broad aponeurosis, inserted into ensiform cartilage, linea alba, os pubis, Poupart's ligament. It is also inserted into anterior superior spinous process of ilium, and outer edge of anterior half of crista ilii; a triangular opening, formed by the separation of the aponeurotic fibres, the inner passing to the symphysis pubis, the outer to the spine of pubes, is called the external abdominal ring.

Obliquus internus vel ascendens. *O.* fascia lumborum, anterior two-thirds of crista ilii, and external half of Poupart's ligament. *I.* cartilages of five inferior ribs, ensiform cartilage, linea alba, also by conjoined tendons into upper edge of pubes, and into linea ileo-pectinea.

Cremaster. This muscle, formed by the lower fibres of internal oblique, is here described, though being a muscle of the testicle. *O.* inner surface of external third of Poupart's ligament, and from lower edge of obliquus internus, and also from transversalis; the fibres pass through the external abdominal ring, forming loops, both in front and behind the spermatic cord. *I.* crest of the pubes.

Transversalis. *O.* fascia lumborum, anterior three-fourths of crista ilii, iliac third of Poupart's ligament, and inner side of six or seven inferior ribs. *I.* along with posterior lamina of obliquus internus, into the whole length of linea alba, upper edge of pubes, and the linea ileo-pectinea.

Rectus. *O.* upper and anterior part of pubes. *I.* ensiform cartilage, costo-xiphoid ligament, and cartilages of fifth, sixth, and seventh ribs.

Pyramidalis. *O.* broad from pubes. *I.* linea alba, midway to umbilicus; sometimes wanting.

DEEP MUSCLES OF THE ABDOMEN.

Diaphragm, is a thin muscular and aponeurotic septum between the chest and the abdomen. *O.* posterior surface of xiphoid cartilage, internal surfaces of cartilages of the last true and all the false ribs, external or false ligamentum arcuatum, and convex edge of true ligamentum arcuatum. *I.* cordiform tendon.

Crura of diaphragm. *O.* right crus from fore part of bodies of second, third, and fourth lumbar vertebræ. Left crus from the sides of the second and third lumbar vertebræ. *I.* posterior border of cordiform tendon. There is a foramen in the diaphragm, for the passage of the vena cava, one for the œsophagus, and another for the aorta, thoracic duct and vena azygos. It also transmits the great sympathetic and the splanchnic nerves.

Quadratus lumborum. *O.* posterior fifth of spine of ilium, and from ilio-lumbar ligament. *I.* extremity of transverse processes of four superior lumbar vertebræ, also inner surface of posterior half of last rib.

Psoas parvus. *O.* side of last dorsal and first lumbar vertebræ. *I.* linea ileo-pectinea, fascia iliaca, and fascial lata, behind the femoral vessels; sometimes wanting.

Psoas magnus. O. sides of bodies of last dorsal, and from bodies and transverse processes of all the lumbar vertebræ; also from inter-vertebral ligaments. I. inferior part of lesser trochanter and ridge below that process.

Iliacus internus. O. iliac fossa, with inner margin of the crista; two anterior spines of ilium, and intervening notch, base of the sacrum, ilio-lumbar ligament, and capsula of hip-joint. I. common tendon with psoas magnus; the inferior fibres are inserted into anterior and inner surface of femur, below trochanter minor. This muscle, with the lower part of the psoas magnus, is bound down by the fascia iliaca.

MUSCLES OF MALE PERINÆUM.

Sphincter ani. O. ano coccygeal ligament. I. into raphe, superficial fascia, and common central point of perinæum.

Sphincter internus encircles the lower part of the rectum.

Erector penis. O. inner surface of tuber ischii, and from insertion of great or inferior sciatic ligament. I. fibrous membrane of crus penis.

Accelerator urinæ. O. central tendon and raphe of perinæum. I. posterior fibers into ramus of pubes; anterior fibres into body of penis; middle fibres surround the bulb and neighboring corpus spongiosum urethræ.

Transversalis perinæi. O. inside of tuber ischii. I. central point of perinæum.

Levator ani. O. posterior part of body of pubes, spine of the ischium, and between these points the angle of division of deep pelvic fascia into obdurator and vesical. I. anterior fibres into central point of perinæum and prostate; middle fibres into side of rectum, posterior fibres into back part of rectum and os coccygis.

Compressor urethræ. O. From the posterior surface of the pubes, near its lower edge and a little on each side of the symphysis. I. below membranous portion of urethra, into the narrow tendinous line, which becomes lost in central point of perinæum.*

Coccygeus. O. inner surface of spine of ischium. I. extremity of sacrum and side of coccyx.

MUSCLES OF FEMALE PERINÆUM.

Sphincter ani, as in the male.

Levator ani. O. as in the male. I. it forms a loop round the vagina as well as around the urethra and rectum.

Coccygeus, as in the male.

---

* This muscle is more correctly described as arising by a narrow tendon from the radius of the pubes on either side, and as terminating in fleshy fibres which decussate in the mesial line, some passing above, others passing below, the membranous portion of the urethra.

Transversalis perinæi, as in the male.

Erector clitoridis. *O.* inner surface of tuber ischii, and insertion of great sciatic ligament. *I.* fibrous membrane of crus clitoridis.

Sphincter vaginæ represents the accelerator urinæ in the male, and extends from the clitoris superiorly, around each side of vagina, to central point of perinæum, in front of anus.

MUSCLES OF THE INFERIOR EXTREMITY.

Fore Part and Sides of the Thigh.

Tensor vaginæ femoris. *O.* external part of anterior superior spine of ilium. *I.* fascia lata about three or four inches below the great trochanter.

Sartorius. *O.* anterior superior spine of ilium and notch beneath it. *I.* inner side of upper end of tibia, below its tubercle.

Rectus femoris. *O.* by two strong tendinous heads: the outer from rim of acetabulum; the inner from anterior inferior spine of ilium. *I.* upper edge of patella.

Vastus externus. *O.* root and anterior part of great trochanter, outer edge of linea aspera, oblique ridge leading to external condyle, external surface of femur, and fascia lata. *I.* external edge of tendon of rectus, side of patella.

Vastus internus. *O.* root of small trochanter and line leading from it to linea aspera, anterior part of femur, inner edge of linea aspera, and ridge leading to inner condyle. *I.* inner edge of tendon of rectus and patella.

Crureus. *O.* anterior and external part of femur, commencing at linea inter-trochanterica, and extending along three-fourths of the bone, as far outwards as linea aspera. *I.* upper and outer edge of patella.

The rectus, crureus, and two vasti, are sometimes described as one muscle, named quadriceps extensor cruris.

Subcrureus or capsularis. *O.* inferior fourth of anterior surface of femur. *I.* Synovial membrane of knee-joint.

Gracilis. *O.* lower part of body and ramus of pubes, and part of ascending ramus of ischium, from a surface of two inches and a half in length. *I.* superior part of internal surface of tibia, just below its tubercle.

Pectineus. *O.* pectineal line external to spine of pubes, and smooth surface of bone in front. *I.* rough ridge leading from lesser trochanter to linea aspera.

Triceps adductor femoris.

1st, *adductor longus.* *O.* anterior surface of pubes, between spine and symphysis. *I.* middle third of linea aspera.

2d, *adductor brevis.* *O.* anterior inferior surface of pubes, between symphysis and thyroid foramen. *I.* superior third of internal root of linea aspera to three inches below lesser trochanter.

3d, *adductor magnus.* *O.* anterior surface of descending ramus of pubes, ramus of ischium, and external border of its tuberosity. *I.* rough ridge leading from great trochanter to linea aspera, linea aspera, and internal condyle of femur.

MUSCLES OF HIP.

Gluteus maximus. *O.* posterior fifth of crista ilii, the rough surface between it, and superior semi-circular ridge, posterior ilio-sacral ligaments, lumbar fascia, spines of sacrum, side of coccyx, and great sciatic ligament. *I.* rough ridge leading from great trochanter to linea aspera, upper third of linea aspera, and fascia lata.

Gluteus medius. *O.* deep surface of fascia covering it, anterior three-fourths of crista ilii, superior semi-circular ridge, and surface of ilium, above and below it. *I.* upper and outer part of great trochanter.

Gluteus minimus. *O.* inferior semi-circular ridge on dorsum of ilium, rough surface between it and edge of acetabulum. *I.* upper and anterior part of great trochanter.

Pyriformis. *O.* concave aspect of second, third, and fourth divisions of sacrum, upper and back part of ilium, and anterior surface of great sciatic ligament. *I.* upper part of great trochanter, behind the insertion of gluteus minimus.

Gamellus superior. O. spine of ischium. I. root of great trochanter.

Gamellus inferior. O. upper part of tuber ischii, and great sciatic ligament. I. root of great trochanter.

Obturator internus. O. pelvic surface of obturator ligament, circumference of obturator foramen, and obturator fascia. I. root of great trochanter.

Quadratus femoris. O. external surface of tuber ischii. I. linea quadrata.

Obturator externus. O. inferior surface of obturator ligament and surrounding surfaces of pubes and ischium. I. root of great trochanter.

MUSCLES ON BACK PART OF THIGH.

Biceps flexor cruris. Long head. O. outer and back part of tuber ischii, by a tendon common to it and to the semitendinosus. *Short head.* O. linea aspera, from below insertion of gluteus maximus to within two inches of external condyle. I. head of fibula.

Semitendinosus. O. tuberosity of ischium, and from three inches of the tendon of the long head of biceps. I. inner surface of tibia, below tubercle.

Semimembranosus. O. upper and outer part of tuber ischii. I. inserted by three portions: 1*st*, into head of tibia, the tendon passing under internal lateral ligament; 2*d*, into the fascia, covering the popliteus muscle; 3*d*, into external condyle of femur, crossing the knee-joint, and forming the ligamentum posticum Winslowii.

MUSCLES ON ANTERIOR AND EXTERNAL PART OF LEG.

Tibialis anticus. O. outer part of two superior thirds of tibia, inner half of interosseous ligament, fascia of leg, and intermuscular septa. I. inner side of internal cuneiform bone, and base of first metatarsal bone.

Extensor digitorum longus. O. external part of head of tibia, head and three-fourths of fibula, part of interosseous ligament, fascia of leg, and intermuscular septa. I. last phalanges of four outer toes.

Extensor pollicis proprius. O. inner edge of middle third of fibula, interosseous ligament, and lower part of tibia. I. base of second phalanx of great toe.

Peroneus tertius. O. anterior surface of lower third of fibula. I. base of fifth metatarsal bone.

Extensor digitorum brevis. O. upper and anterior part of os calcis, calcaneo-astragaloid, and annular ligaments. I. internal tendon into base of first phalanx of great toe; the three others join the outer edges of corresponding tendons of extensor digitorum longus.

MUSCLES ON OUTER PART OF LEG.

Peroneus longus. *O.* upper two-thirds of outer surface of fibula, small portion of tuberosity of tibia, fascia of leg, and inter-muscular septa. *I.* tendon passes in a groove in os cuboides obliquely across the sole of the foot, to become attached to the tarsal end of metatarsal bone of great toe.

Peroneus brevis. *O.* outer and back part of lower half of fibula and intermuscular septa. *I.* base of metatarsal bone of little toe.

MUSCLES ON BACK OF LEG.

Superficial layer.

Gastrocnemius. *O. internal head,* upper and back part of internal condyle of femur, and oblique ridge above it; *external head,* from above external condyle. *I.* lower and back part of os calcis.

Plantaris. *O.* back part of femur above external condyle, and posterior ligament of knee. *I.* posterior part of os calcis.

Soleus. *O. external head,* from back part of head and superior third of fibula; *internal head,* from middle third of tibia, below insertion of popliteus, and from a tendinous arch extending between the bones over the posterior tibial vessels: unites with gastrocnemius to form tendo Achillis. *I.* lower and back part of os calcis.

Popliteus. O. depression on outer condyle of femur. I. flat triangular surface, occupying the superior posterior fifth of tibia.

Flexor digitorum longus. O. posterior flat surface of tibia, from below popliteus to within three inches of ankle, fascia, and intermuscular septa. I. last phalanges of four lesser toes.

Tibialis posticus. O. posterior internal part of fibula, upper part of tibia, and nearly whole length of interosseous ligament. I. inferior and internal tuberosity on os naviculare, internal cuneiform and cuboid bones, and second and third metatarsal bones.

Flexor pollicis longus. O. two inferior thirds of fibula. I. last phalanx of great toe.

MUSCLES OF FOOT.

First Layer.

Abductor pollicis. O. lower and inner part of os calcis, internal annular ligament, plantar aponeurosis, and internal intermuscular septum. I. internal side of base of first phalanx of great toe; there is a sesamoid bone in the tendon.

Flexor digitorum brevis perforatus. O. inferior and internal part of os calcis, plantar aponeurosis, and intermuscular septa. I. second phalanges of four lesser toes.

Abductor minimi digiti. O. outer side of os

calcis, ligament extending from os calcis to outer side of fifth metatarsal bone, plantar fascia, external intermuscular septum, and base of fifth metatarsal bone. *I.* outer side of base of first phalanx of little toe.

Second Layer.

Musculus accessorius. *O.* inferior and internal part of os calcis. *I.* outer part of tendon of flexor digitorum longus.

Lumbricales. *O.* tendons of flexor digitorum longus. *I.* internal side of first phalanges of four lesser toes. Between the flexor brevis and flexor accessorius lie the external plantar vessels and nerve.

Third Layer.

Flexor pollicis brevis. *O.* by a pointed process from inner border of os cuboides and from external cuneiform bone. *I.* by two divisions into outer and inner border of base of first phalanx of great toe : the tendons contain sesamoid bones.

Abductor pollicis. *O.* os cuboides, base of third and fourth metatarsal bones and sheath of peroneus longus. *I.* base of first phalanx of great toe.

Transversalis pedis. A narrow fasciculus stretched beneath the digital extremities of the metatarsal bones.

Flexor brevis minimi digiti. *O.* fifth metatarsal bones and sheath of tendon of peroneus longus. *I.* inner side of base of first phalanx of little toe.

Fourth Layer.

Seven interossei muscles. Three on sole of foot, and four upon its dorsum.

Inferiores, vel plantares.

1st. *O.* inner side of third metatarsal bone. *I.* base of first phalanx of the same toe.

2nd. *O.* inner side of fourth metatarsal bone. *I.* inner side of first phalanx of the same toe.

3rd. *O.* fifth metatarsal bone. I. inner side of base of first phalanx of little toe.

Superiores, vel dorsales; they arise by two heads from the contiguous surfaces of the metatarsal bones.

1st. *O.* internal side of second metatarsal bone and outer side of first. *I.* inner side of base of first phalanx of second toe.

2nd. *O.* opposite sides of second and third metatarsal bones. *I.* outer side of first phalanx of second toe.

3rd. *O.* opposite side of third and fourth metatarsal bones. *I.* outer side of first phalanx of middle toe.

4th. *O.* opposite sides of the fourth and fifth metatarsal bones. *I.* outer side of the first phalanx of fourth toe.

THE MUSCLES OF THE ORBIT.

1. *Levator palpebræ superioris.* *O.* upper

edge of the foramen opticum. *I.* superior border of tarsal cartilage.

2. *Obliquus superior.* *O.* foramen opticum. *I.* sclerotic coat between superior and external rectus.

3. *Obliquus inferior.* *O.* orbital edge of superior maxillary bone. *I.* sclerotic coat between it and external rectus muscle.

4. *Rectus superior.*
5. *Rectus inferior.*
6. *Rectus internus.*
7. *Rectus externus.*

[All arise round optic foramen, the external rectus being also attached to margin of sphenoidal fissure, near the origin of the superior rectus, and they are inserted about a quarter of an inch behind cornea. The internal rectus is the muscle most frequently divided in the operation for strabismus.]

The third, fourth, nasal division of fifth, sixth nerves, and the ophthalmic vein, pass between the two heads of the rectus externus.

MUSCLES OF INTERNAL EAR.

Stapedius. *O.* within pyramid. *I.* neck of stapes.

Tensor Tympani. *O.* canal in petrous portion of temporal bone, above Eustachian tube. *I.* short process below neck of malleus.

Laxator Tympani. *O.* spinous process of sphenoid bone and Eustachian tube. *I.* processus gracilis of malleus.

BRAIN AND ITS MEMBRANES.

DURA MATER.

A firm, dense, fibro-serous membrane, adhering by its outer surface to the bones of the cranium, its inner surface being intimately connected with the arachnoid membrane. It defends the brain, acts as an internal periosteum to the bones of the skull, forms the sinuses, and sends envelopes upon the several nerves as they pass through the cranial holes. It sends off the following processes.

Falx cerebri commences narrow at crista galli, and arches backwards between the lobes of the cerebrum, becoming deeper until it meets the tentorium, with which process it is continuous on either side. Its convex edge corresponds to the median groove of the os frontis, the sagittal edges of the parietal bones, and the upper half of the perpendicular ridge of the occipital bone. The great longitudinal sinus is in its upper edge, and the lesser longitudinal sinus in its inferior free concave edge.

Tentorium cerebelli extends in a horizontal manner above the cerebellum and below the posterior lobes of the cerebrum. Its convex edges contain the lateral sinuses, and correspond to the transverse ridges of the occipital bone, the inferior posterior angles of the parietal bones, the superior angles of the petrous portion of the temporal bones, and to the clinoid processes of the sphenoid bone. It separates the cerebrum from the cerebellum, and is bony in the carnivorous animals.

Falx cerebelli is attached to the lower half of the perpendicular ridge of the occipital bone, and extends between the lobes of the cerebellum towards the foramen magnum.

Sphenoidal folds are attached to the lesser wings of the sphenoid bone.

SINUSES.

Great longitudinal sinus, of triangular form, extends along the convex margin of the falx cerebri. It commences by a small vein in the foramen cœcum, and increasing in size as it proceeds backwards, pours its blood into the torcular Herophili. Its interior is crossed by small bands called *chordæ Willisii*, and presents the openings of the veins which course upon the upper surface of the cerebral hemispheres, and a number of small whitish granules called *glandulæ Pacchioni*. They are also found upon the outer surface of the upper wall of the sinus between it and the cranium.

Inferior longitudinal sinus is very small, runs along the concave edge of the falx cerebri, and terminates in the straight sinus.

Straight sinus passes from the termination of the inferior longitudinal sinus downwards and backwards, receiving the blood of the venæ

Galeni, and empties itself into the torcular Herophili.

Lateral sinuses, each corresponds to the transverse groove in the occipital bone, the groove in the posterior inferior angle of the parietal bone, the mastoid fossa of the temporal bone, and the groove in the occipital bone on either side of the foramen magnum; it passes through the foramen lacerum posterius and becomes the jugular vein.

Torcular Herophili corresponds to the centre of the crucial spine of the occipital bone; six sinuses communicate with it, viz., the two lateral, the great longitudinal, the straight, and the two occipital.

Cavernous sinuses, each extends from the anterior clinoid process to the petrous portion of the occipital bone; and, upon being cut into, presents a cellular appearance. The internal carotid artery, the sixth nerve, and branches of the sympathetic nerve, are found within each, but separated from the blood by the reflected venous lining membrane. In the outer wall of each run the third and fourth nerves, and the first branch of the fifth; the sinus of either side presents the openings of the ophthalmic vein, of the two petrosal sinuses, and of the circular sinus.

Circular sinus surrounds the pituitary body, and is formed of an anterior and posterior transverse vein which extends from one cavernous sinus to the other.

Superior petrosal sinuses, each passes from the cavernous sinus along the upper angle of the petrous portion of the temporal bone, to the lateral sinus.

Inferior petrosal sinuses, each passes from the cavernous sinus downwards and backwards, along the line of contact of the petrous portion of the temporal bone, and the occipital to the lateral sinus, just where this terminates in the internal jugular vein.

Transverse sinus crosses the basilar process of the occipital bone, and connects the inferior petrosal sinuses.

Occipital sinuses, two in number, are contained in the falx cerebelli, and open into the torcular Herophili.

TUNICA ARACHNOIDEA.

Belongs to the class of serous membranes, is spread over the surface of the brain without penetrating between its convolutions, and is reflected upon the dura mater, in those situations where the nerves and veins pierce this fibrous membrane; thus, after the manner of all serous membranes, it forms a shut sac, and consists of a parietal and a visceral layer. It is stated to gain access to the interior of the brain by the great transverse fissure, and to line the free surfaces of the ventricles.

PIA MATER.

The vascular covering of the brain lines its entire surface, dipping between its convolutions and sending numerous blood-vessels into its substance; it is intimately connected to the arachnoid membrane by its outer surface, except at the base and sulci of brain; and entering the ventricles by the great transverse fissure, gives them a lining membrane, and forms the choroid plexuses.

THE BRAIN.

The brain is subdivided into three portions, viz., the cerebrum, the cerebellum, and the medulla oblongata.

THE CEREBRUM.

This is the largest of the three divisions, is of oval form and divided into two equal portions, called hemispheres, by a fissure (superior longitudinal) which extends along the median line upon its proper surface, and contains the falx cerebri and the arteries of the corpus callosum. At the base of the brain the hemispheres are divided at each extremity by this fissure, but in the centre they are united.

Hemispheres, right and left, are convex superiorly and externally, and flat towards each other, where they correspond to the falx.

Lobes. Each hemisphere is divided into three lobes upon its under surface; the anterior, which is the smallest, rests upon the roof of the orbit and presents a groove for the lodgment of the olfactory nerve; the middle, prominent and convex, lies in the middle fossa, in the base of the cranium; and the posterior rests upon the tentorium.

Fissura Sylvii separates the anterior from the middle lobe, and corresponds to the lesser wing of the sphenoid bone and its fold of dura mater. The cerebral surface of this fissure is pierced by small arteries, branches of the middle cerebral artery.

Convolutions, or gyri, are eminences longitudinal and rounded, but directed in various ways upon the surface of each hemisphere.

Sulci are the fissures which separate the convolutions from each other, over which the arachnoid membrane passes, but into which the pia mater dips.

Cineritious substance of brain is of a yellowish-gray color, from three to four lines in thickness, soft and very vascular, and for the most part situated upon the outer surface of the brain. It is however found in striæ and masses in the interior of the brain, and surrounded by the medullary substance. In some situations its color assumes a dark hue, as is seen when a section of the crus cerebri is made. Microscopic examination shows that it is composed of vesicles, containing nuclei and granules: hence it is sometimes called "vesicular matter." This gray vesicular or nucleated cell material is believed to have the power of generating or producing nervous influence. while the white fibrous or tubular substance merely acts as a conductor.

Medullary substance, white and fibrous, forms the greater part of the brain.

Centrum ovale minus, a term applied to the appearance brought into view by making a section of each hemisphere within a few lines of the corpus callosum; an oval mass of white matter is seen surrounded by the gray cortical substance.

Centrum ovale magnum, a term applied to the mass of medullary structure, which is rendered apparent by slicing both hemispheres on a level with the corpus callosum.

Corpus callosum, unites the hemispheres to each other, is about three inches in length, and presents upon its upper surface the *raphè*, which corresponds to the anterior cerebral arteries, and from which, on either side, pass the connecting transverse fibres of the hemispheres, called lineæ transversæ. It unites by its posterior extremity with the fornix and the hippocampus, major and minor; its anterior extremity being curved upon itself and continuous with the optic commissure and tuber cinereum at the base of the brain.

Septum lucidum descends from the raphe of the

corpus callosum to the fornix, separating the lateral ventricles from each other. It consists of two layers, which are composed of white and of gray matter, and lined by epithelium; the cavity between them is called the *fifth ventricle:* its form is triangular, the apex corresponding to the union of the corpus callosum and the fornix, the base anterior, corresponding to the curved portion of the corpus callosum.

Lateral ventricles, each consists of a body and three cornua, the body corresponding to the centre of each cerebral hemisphere, the cornua proceeding one to each lobe. The bodies of the ventricles are separated from each other by the septum lucidum.

Corpora striata, two pear-shaped bodies, their large bulbous extremities being contained in the anterior cornua of the lateral ventricles, their narrow stalk-like extremities being directed backwards into the bodies of the ventricles; they are cineritious on their surface, but when cut into present alternating striæ of cineritious and medullary matter; and hence their name.

Optic thalami, two large bodies placed behind and between the corpora striata; each presents upon its superior surface two tubercles, called *corpora geniculata*. Towards the median line the optic thalami are flat and united to each other by a soft cineritious structure, called *commissura mollis;* upon their external surface they are white, but their interior is gray. The corpora striata and optic thalami, like the cineritious surface of the brain, are very vascular, hence all these parts are apt to be the most frequent seats of apoplectic effusions.

Tænia semicircularis, a narrow medullary band, situated in the groove between the optic thalamus and corpus striatum of either side.

Fornix, a longitudinal commissure, placed horizontally beneath the septum lucidum and corpus callosum, and composed of medullary structure, arches above the third ventricle, and lies upon the velum interpositum and choroid plexus. It commences by its *two posterior crura*, which arise from the hippocampi majores in the inferior cornua of the lateral ventricles; these unite in the median line, and form what is called the *body* of the fornix, which passes forwards and terminates in the *anterior crura:* finally, the *anterior crura* descend to the base of the brain, and terminate in the corpora mammillaria, or albicantia.

The lyra, is the appearance presented upon the under surface of the fornix by some white fibres which connect, by a transverse commissure, the hippocampi and the posterior crura of the fornix.

Choroid plexus, the fold of pia mater which lies upon the optic thalamus, and which enters the body of the lateral ventricle by the inferior cornu; the choroid plexus of either side passes forwards and inwards, and both unite in the foramen commune anterius.

Velum interpositum, lying underneath the fornix, unites the choroid plexuses of either side; it is composed of arachnoid membrane and pia mater, and contains in its centre the venæ Galeni.

Venæ Galeni, contained in the velum interpositum, pass from before backwards and terminate in the straight sinus. These veins return the blood from the choroid plexuses and from the parts within the ventricles.

Pineal body, a small conical cineritious mass containing in general sandy matter, which has been found to consist of phosphate and carbonate of lime. It is placed upon the corpora quadrigemina, and is connected with the optic thalami by two peduncles.

Hippocampus minor, an oval eminence in the posterior cornu of the lateral ventricle; medullary externally, and cineritious in its interior.

Hippocampus major, a similar eminence to the minor, but larger, and placed in the inferior cornu of the lateral ventricle.

Pes hippocampi, the tuberculated appearance which the extremity of the hippocampus major presents.

Tænia hippocampi, or *corpus fimbriatum*, the free margin of the posterior crus of the fornix, where it is connected with the hippocampus major.

Corpus denticulatum, a cineritious serrated line in the inferior cornu of the lateral ventricle, and which is exposed upon removing the tænia hippocampi, beneath which it lies.

Third ventricle, a deep fissure between the optic thalami, exposed by separating these bodies. It is bounded anteriorly by the descending crura of the fornix and the anterior commissure, posteriorly by the posterior commissure and the tubercula quadrigemina, laterally by the optic thalami: its floor corresponds to the locus perforatus, tuber cinerium and infundibulum; it is covered in by the velum interpositum and fornix.

Foramen commune anterius, a hole by which the choroid plexuses unite anteriorly. It forms a medium of communication for the two lateral and the third ventricle.

Infundibulum, a hollow funnel-shaped process of gray matter leading from the anterior and inferior part of the third ventricle to the pituitary body.

Iter e tertio ad quartum ventriculum, an opening in the posterior part of the third ventricle, under the posterior commissure and tubercula quadrigemina, and leading obliquely backwards and downwards to the fourth ventricle.

Anterior commissure, a medullary round cord, anterior to the crura of the fornix, and passing transversely from one corpus striatum to the other.

Posterior commissure, extends transversely from one optic thalamus to the other. It is shorter and smaller than the anterior commissure.

Tubercula quadrigemina, four eminences, called also nates and testes, situated under the posterior part of the velum interpositum; the two anterior (the nates) are connected to the optic thalami; the posterior (the testes) are connected to the cerebellum by the following processes.

Processus e cerebello ad testes, two medullary bands or plates, which pass obliquely from the cerebellum upwards and inwards to the testes.

Valve of Vieussens, a layer of cineritious and medullary substance, of triangular form, attached by its sides to the processus e cerebello ad testes, by its base to the cerebellum, and by its apex to the testes.

Fourth ventricle, is exposed by cutting through the valve of Vieussens. It is bounded anteriorly by the pons Varolii, laterally by the processes e cerebello ad testes, superiorly by the valve of Vieussens, inferiorly by pia mater and arachnoid membrane, and posteriorly by the cerebellum.

Calamus scriptorius, the fissure seen upon the posterior surface of the pons Varolii, in the fourth ventricle, from either side of which four or five white lines proceed.

Choroid plexus of fourth ventricle, a small fold of pia mater, which enters the ventricle as this membrane is passing from the cerebellum to the spinal cord.

CEREBELLUM,

Consists of a central portion called superior and inferior vermiform processes, and of two hemispheres, united inferiorly by the pons Varolii.

Hemispheres, are flat superiorly, where they correspond to the tentorium, and convex inferiorly, where they lie in the inferior occipital fossa: the surface of each presents semicircular narrow lines, arising from the laminated arrangement of the cineritious portion of the organ; between these laminæ the pia mater enters, but the arachnoid passes over them.

Crura cerebelli, two medullary cords which pass from either hemisphere and unite in the pons Varolii.

Superior vermiform process, a small conical eminence corresponding to the superior and central part of the cerebellum.

Inferior vermiform process, larger than the superior, and corresponding to the inferior and central part of the cerebellum.

Ar or vitæ, the branching of the medullary substance of the cerebellum, exposed by making a vertical section of it.

Corpus dentatum, a small oval mass of cineritious substance, surrounded by a yellow zigzag line, and exposed upon making a section of the cerebellum, parallel to, but an inch distant from, the median line.

MEDULLA OBLONGATA,

A large conical process of medullary structure,

extending from the lower margin of the pons Varolii to the commencement of the spinal cord. It is rather more than an inch in length, and presents the following bodies, which are separated from each other by distinct grooves.

Corpora pyramidalia, or the anterior pyramids, the two anterior eminences of the medulla oblongata.

Corpora olivaria, smaller than the pyramidal bodies, situated laterally.

Corpora restiformia, large, and situated posteriorly; *the posterior pyramids*, small and cordlike, lie within the restiform bodies upon the posterior surface of the cord.

BASE OF THE BRAIN

Presents on either side of the median line the anterior and middle lobes of the cerebrum, separated from each other by the fissure of Sylvius, and a lobe of the cerebellum resting upon the posterior lobe of the cerebrum. In the median line, proceeding from before backwards, is the anterior extremity of the median fissure (on either side of which run the olfactory nerves), the lower extremity of the corpus callosum, the optic commissure, the tuber cinereum, the corpora albicantia, the pituitary body and infundibulum, the locus perforatus (on either side of this is the crus cerebri), the pons Varolii, the medulla oblongata, and lastly the posterior extremity of the median fissure.

The brain consists—

1. Of certain masses of grey, or cineritious matter, microscopically consisting of nucleated cells, and having white or tubular matter intermixed or passing through it. To these masses the name of "ganglia" has been applied. Examples: the cortical surface of the brain and cerebellum, the corpora striata, optic thalami, locus niger, etc., etc.

2. Of commissures, or bands of union, connecting these ganglia and other parts. Thus, the corpus callosum unites the hemispheres, the anterior commissure, the corpora striata, the middle and posterior unite the optic thalami, the pons Varolii unites the two lobes of the cerebellum, etc., etc. Many of these commissures are formed of *converging fibres.*

3. Of diverging fibres. These can be traced upwards from the columns of the spinal cord into the medulla oblongata, where they form the corpora pyramidalia, olivaria, and restiformia, etc. The corpora pyramidalia are continuous with the antero-lateral columns of the spinal cord, from which motor nerves arise; they can be traced up under the pons Varolii, expanding always as they pass through grey matter, passing on to help form the crura cerebri, still expanding they pass through the optic thalami and corpora striata, until they spread out in the convolutions of the brain. Motor nerves arise from the corpora pyramidalia, from its continued track, as traced upwards. The corpora pyra-

midalia decussate below the pons Varolii. This explains the fact that when pressure is made upon one hemisphere of the brain, as by depressed bone, etc., the paralysis will generally be found to affect the opposite side of the body.

ORIGIN OF THE CEREBRAL NERVES.

First nerve (olfactory), arises by three roots, the external, long, and white, from the fissure of Sylvius; the internal, also white, from the posterior internal surface of the under part of the anterior lobe; and the middle short and cineritious, from a grey tubercle upon the under surface of the anterior lobe.

Second pair (optic), arise by two roots from the corresponding nates and from the corpora geniculata; the roots unite and form the tractus opticus, which passes around and becomes slightly attached to the crus cerebri; the tracts, one from either side, then unite in the optic commissure, having previously received a few fibres from the tuber cinereum; from the anterior part of this commissure proceed the optic nerves.

Third pair (motores oculorum), arise from the inner side of the crus cerebri, near the pons Varolii.

Fourth pair (trochleares), arise from the valve of Vieussens, by several delicate filaments, which meet those of the opposite side in the mesial line.

Fifth nerve (trigeminal), consists of two portions, one for sensation, the other for motion.

The motor division, the smallest, arises from the corpus pyramidale, in the substance of the pons Varolii; and the *sensory division*, from the angles between the crus cerebelli and the pons Varolii.

Sixth pair (abducentes), arise from the corpus pyramidale near its junction with the pons Varolii.

Seventh nerve consists of the *portio dura*, or facial, and the *portio mollis* or auditory. The *portio dura* arises from the groove between the corpus restiforme and olivare, near the pons Varolii. The *portio mollis* arises by three or four white lines from the calamus scriptorius in the fourth ventricle.

Eighth nerve consists of the *glosso-pharyngeal*, *pneumo-gastric* and *spinal accessory*. The *glosso-pharyngeal* arises from the groove between corpus olivare and corpus restiforme, by four or five filaments. The *pneumo-gastric* arises in the same groove, but below the glosso-pharyngeal, by eight or ten filaments, and the *spinal accessory* arises from the side of the spinal cord, as low as the sixth cervical vertebra, by several filaments.

The ninth pair (linguales) arise by seven or eight filaments from the groove between the pyramidal and olivary bodies, about half an inch below the origin of the sixth.

DISTRIBUTION OF CEREBRAL NERVES.

First pair (or olfactory), sends off three sets of branches to the upper part of the nose. *In-*

ternal branches to septum nasi; *middle branches* to mucous membrane of roof of each nostril; and *external branches* to spongy bones.

Second pair (or optic), pierce the sclerotic coat of the eye, and form the retina.

Third pair (or motores oculorum). *Superior, or smaller branch*, supplies the superior rectus, and the levator palpebræ. *Inferior, or larger branch*, supplies the internal rectus, the inferior rectus, and the inferior oblique; and also sends a branch to the lenticular ganglion.

Fourth pair (or trochleares), are distributed to superior oblique muscles of eye.

Fifth pair (or trigemini), first form the Casserian ganglion, and divide into three main branches—viz., ophthalmic, superior and nferior maxillary.

Ophthalmic division.—Divides into—1. Lachrymal nerve, which, passing along the outer wall of the orbit, sends a branch downwards to the orbital branch of the superior maxillary nerve; then gives filaments to the lachrymal gland and the conjunctiva; and finally, perforating the fibrous attachment of the upper eyelid, terminates in the integument of the forehead. 2. Frontal nerve, enters the orbit above the levator palpebræ; it divides into supra-orbital and supra-trochlear: supra-orbital nerve, escaping through the supra-orbital notch, is distributed to the integuments of the forehead; the supra-trochlear, passing above the pulley of the trochlearis muscle, sends a filament to the infra-trochlear branch of the nasal nerve, and terminates in the mucous membrane of the inner canthus and in the integuments of the forehead. 3. Nasal nerve, enters the orbit between the two heads of the rectus externus muscle. It gives a filament to the lenticular ganglion, two or three ciliary nerves, and the infra-trochlear branch; the nerve then enters the skull by the foramen ethmoideum anterius, and escaping into the nose by a fissure in the cribriform plate of the ethmoid bone, terminates at the tip of the nose: the terminal filament is called naso-lobular.

Superior maxillary division.—1. Orbital branch, which sends off a malar twig and a temporal twig. 2. Two branches to Meckel's ganglion. 3. Posterior dental, which sends off an anterior branch to the buccinator muscle and gums, and a posterior branch to the molar teeth. 4. Anterior dental, to the antrum and teeth. 5. Infra-orbital, distributed to the face.

Inferior maxillary division.—1. The superior or external branch, which is joined by the motor portion of the fifth nerve, sends off deep temporal twigs—a masseteric branch to masseter muscle and temporo-maxillary articulation, a buccal branch to the buccinator and temporal muscles, and a pterygoid branch to the pterygoid and palatine muscles. 2. Inferior or internal branch, sends off the inferior dental (which gives off the mylohyoid nerve), it supplies the teeth, and terminates in the mental nerve; the gustatory, which goes to the glands, mucous membrane, and papillæ of the tongue, and is joined by the chorda tympani nerve; and the auriculo-temporal, which supplies the external ear, and the integuments of the side of the head.

GANGLIONS IN CONNEXION WITH THE FIFTH PAIR.

Casserian ganglion.—A large grey semi-lunar body, analogous to the ganglion upon the posterior roots of the spinal nerves. It lies in a depression at the end of the petrous portion of the temporal bone, and presents anteriorly a convex border, from which proceed the three main divisions of the fifth, just described. The motor root of the fifth nerve joins the inferior maxillary division only.

The lenticular ganglion is situated between the optic nerve and external rectus muscle; it receives at its posterior superior angle a long filament from the nasal branch of the fifth, and by its posterior inferior angle a branch from the inferior division of the third. Its anterior angles furnish the ciliary nerves, about twenty in number, which run along the optic nerve, pierce the back part of the sclerotica, run forward between it and the choroid coat, enter the ciliary ligament, and are ultimately distributed to the iris.

Meckel's ganglion.—A little, red body of triangular shape, situated deep in the fat and cellular tissue of the pterygo-maxillary fossa; it communicates superiorly by two small nervous twigs with the second division of the fifth, and sends off—1. Spheno-palatine nerve to the mucous membrane of the superior and middle spongy bones; it gives off the naso-palatine branch which runs along the septum nasi, and terminates in the foramen incisivum. 2. Palatine nerve, which descends in the palatine canal, and divides into anterior branches which supply the teeth, and posterior and middle branches which supply the amygdalæ, soft palate, and uvula. 3. Vidian nerve, which passes backwards through the Vidian canal, enters the cranium by the foramen lacerum anterius, and divides into an inferior and a superior branch, having first sent filaments to the sphenoidal sinus: the inferior branch enters the carotid canal, and unites with the branches of the sympathetic, whilst the superior branch runs beneath the Casserian ganglion on the petrous portion of the temporal bone, enters the hiatus Fallopii, attaches itself to the portio dura nerve, leaves it, and enters the tympanum a little below the pyramid, and is here called chorda tympani. It then passes between the long crus of the incus and the handle of the malleus, emerges from the tympanum by the Glasserian fissure, unites with the gustatory nerve, and at the submaxillary gland terminates in a small ganglion named after Boch.

Otic ganglion, a small body connected with the inferior maxillary nerve, near the foramen ovale; it receives the lesser petrosal nerve of the glosso-pharyngeal.

The naso-palatine ganglion lies in the anterior

palatine hole, and is formed by the anterior palatine branches from Meckel's ganglion.

The submaxillary ganglion lies at the edge of the submaxillary gland, and is formed by the termination of the chorda tympani nerve.

Sixth pair, or abducentes, are distributed to the external rectus muscle of each side exclusively, but receive two filaments from the sympathetic in the cavernous sinus.

Seventh pair consists of two portions, viz., portio dura and portio mollis; both enter the meatus auditorius internus.

Portio dura, or facial nerve, passes into the aqueduct of Fallopius, and escapes by the stylo-mastoid foramen. It gives off, 1, a branch to the stapedius muscle; 2, the chorda tympani; 3, posterior auricular; 4, digastric; 5, stylo-hyoid nerves: and then divides into, 1, cervico-facial division, which sends off branches to the muscles of the face and to the platysma myoides; and 2, temporo-facial division, which sends off temporal, malar, and buccal branches. The interlacement of the branches of the facial nerve, as it passes through the parotid gland, is called "pes anserinus."

Portio mollis, or auditory nerve.—1. Branch to cochlea. 2. Branch to vestibule and semicircular canals.

Eighth nerve consists of three portions, viz., glosso-pharyngeal, pneumo-gastric, and spinal accessory. They escape from the skull by the jugular foramen.

Glosso-pharyngeal, or first branch of the eighth, gives off, 1, Jacobson's nerve, which enters the tympanum by a small foramen upon the under surface of the temporal bone, and sends a filament (small petrosal) to the otic ganglion, and carotid filaments which ramify on the coats of the vessel, and communicate with the sympathetic and vagus nerves. 2. Branches to the pharyngeal plexus. 3. Branches to the tonsillitic plexus. 4. Branches to the stylo-pharyngeus and superior and middle constrictors of the pharynx, mucous membrane of fauces, &c. 5. Branches to the papillæ and mucous membrane at the root of the tongue.

Pneumo-gastric, or second branch of the eighth, gives off, 1. Branches to unite with the spinal accessory, glosso-pharyngeal, lingual, and sympathetic nerves. 2. Branches to assist in forming the pharyngeal plexus. 3. Superior laryngeal nerve, which is the nerve of sensation of the larynx, gives off an external laryngeal branch to the exterior of the larynx, the inferior constrictor and pharynx, and crico-thyroid muscle, and then pierces the thyro-hyoid membrane in company with the superior laryngeal artery, and supplies the epiglottis, mucous membrane, and arytenoid muscles. 4. Cardiac branches, to the cardiac nerves of the sympathetic, is the nerve of motion of the larynx. 5. Inferior laryngeal or recurrent nerve, which sends off cardiac filaments, branches to the fore part of the trachea and thyroid gland, and branches to the pharynx, laryngeal muscles, and mucous membrane, on which they communicate with branches of the superior laryngeal. 6. Pulmonary branches, which send off branches in front of the bronchial tubes to form the anterior or lesser pulmonic plexus; this plexus sends filaments to the pulmonary vessels, also to the lungs and pericardium, and to the posterior pulmonic plexus. 7. Posterior, or greater pulmonic plexus, is formed by the pneumo-gastric nerves, which increase in size at the root of each lung, and subdivide and unite in an areolar manner. This plexus is joined by several branches of the sympathetic nerve, and its branches accompany the bronchial tubes through the substance of the lung. 8. Œsophageal plexus, or plexus gulæ, is formed by the communications of both nerves, encircling the œsophagus in their course along this tube. 9. Gastric plexus is formed by both nerves dividing, subdividing, and uniting upon the stomach. The left pneumo-gastric nerve is anterior upon the stomach, and sends branches to the lesser omentum and liver; the right is posterior.

The pneumo-gastric supplies the pharynx, the larynx, the trachea, the bronchial tubes, the œsophagus, the heart, lungs, and stomach.

Nervous accessorius, or third branch of the eighth. 1. Branches to communicate with the eighth, ninth, and sympathetic nerves. 2. Branches to the sterno-cleido-mastoid muscle, which muscle it then perforates. 3. Terminal branches to the trapezius muscle.

Ninth pair, or lingual. 1. Descendens noni unites with the internal descending branches of the cervical plexus, forming a small plexus in loops, the branches of which pass to the omo-hyoid, sterno-hyoid, and sterno-thyroid muscles. 2. A branch to the thyro-hyoid muscle. 3. Branches to the hyoglossus and surrounding muscles, and to the gustatory branch of the fifth pair. 4. Terminal branches to the genio-hyoglossus muscle, and muscular structure of the tongue.

SPINAL NERVES.

Symmetrical, thirty pairs, viz., eight cervical, twelve dorsal, five lumbar, and five sacral. Each spinal nerve has two roots, an anterior and a posterior. The anterior is small, and is the motor division. The posterior large, with a ganglion upon it, and is for sensation. These roots are separated by the cord itself, and by the ligamentum dentatum. The anterior root is connected to the posterior root beyond the ganglion. On the outer side of the ganglion both nerves unite in a single cord, which, after a short course, divides into an anterior and posterior branch. The posterior branches of this division are the smaller (except that of the second cervical), and are distributed to the dorsal muscles and integuments. The anterior branches form the several plexuses which supply the muscles and integuments anterior to the spine, and also the extremities.

DISTRIBUTION OF THE EIGHT CERVICAL NERVES AND FIRST DORSAL NERVES.

Posterior branches are small, except the second cervical, which perforates the complexus muscle, and accompanies the occipital artery; the rest are lost in the neighboring muscles and integuments.

Anterior branches. The first, or sub-occipital, twists round the atlas, to unite with the second, forming the nervous loop of the atlas: and the second, having received the first, descends to unite with the third. The third unites in like manner with the fourth, and thus is formed, by the anterior branches of the four first cervical nerves, the

CERVICAL PLEXUS.

From this plexus proceed:—

Branches to the platysma, integuments, parotid gland, ear, and back of the head.

1. *Great auricular*, arises chiefly from the third cervical, and is distributed to the ear; it accompanies the external jugular vein.

2. *Small occipital*, which, arising from the second cervical, pierces the fascia behind the sterno-mastoid, and is distributed to the integuments of the scalp.

3. *Descending branches* from third and fourth

cervical, which are divided into sternal, clavicular, and acromial, and supply the integuments.

4. *Branches*, generally two in number, which form loops with the descendens noni in front of the jugular vein.

5. *Phrenic, or internal respiratory*, which arises from the third and fourth cervical, and has a small filament also from the fifth cervical: it sends branches to the liver, pericardium, inferior cava, and terminates in the diaphragm. The phrenic is the most important nerve in the human body, as upon it depends the action of the diaphragm.

6. *Muscular branches*, which are given to the sterno-mastoid and trapezius (these muscles are also supplied by the spinal accessory); to the levator anguli scapulæ, the scaleni, and recti capitis antici.

THE BRACHIAL PLEXUS

Is formed by the union of the anterior branches of the four inferior cervical and first dorsal nerves. From the plexus proceed:—

1. *A branch* to join the phrenic nerve.
2. *Branches* to the longus colli, scaleni, and subclavius muscles.
3. *The external respiratory nerve of Bell*, which, arising from the fifth and sixth cervical, passes behind the axillary vessels, and is distributed to the serratus magnus.
4. *Thoracic nerves*, three or four in number,

which form loops round the axillary artery, and supply the pectoral muscles.

5. *Supra-scapular nerve*, which passes through the notch in the scapula, and supplies the supra-spinatus, infra-spinatus, and teres minor muscles.

6. *Subscapular nerves* are three or four in number; they descend behind the vessels to the subscapular, latissimus dorsi, and teres major muscles.

7. *Internal cutaneous nerve*, sends one branch, which descends over the bend of the elbow as low as the wrist, and another branch which descends towards inner condyle, and sends branches to inner and posterior part of the fore-arm.

8. *External cutaneous*, musculo-cutaneous or perforans Casserii, which sends branches to the coraco-brachialis, biceps, and brachialis anticus; branches to the integuments of the fore-arm, an anterior branch to the ball of the thumb and palm of the hand; and a posterior branch to the dorsum of the hand.

9. *Median nerve*, sends branches to the superficial and deep pronators and flexors of the fore-arm, except the flexor carpi ulnaris, and half the flexor digitorum profundus, which are supplied by the ulnar nerve; the anterior interosseous nerve, which sends a branch to the pronator quadratus, and another to the dorsum of the hand; a superficial branch which is given off above the wrist, and which runs to the palm of

the hand; and five digital branches, which supply the thumb, index, and middle fingers, and the radial edge of the ring finger.

10. *Ulnar nerve*, sends muscular branches to skin of fore-arm, flexor profundus, and flexor carpi ulnaris muscles; the nervus dorsalis carpi ulnaris to the integuments on the dorsum of the hand and the three inner fingers; the superficial palmar branch, which divides into three digital branches for the supply of the little finger and the ulnar edge of the ring finger; and the deep palmar branch to form the deep palmar arch, which supplies the interossei muscles, two inner lumbricales, and the adductor pollicis muscles.

11. *Musculo-spiral*, or radial nerve, sends branches to the triceps, through which it winds; a long cutaneous branch to the elbow; branches to the supinators and extensors; the anterior or radial branch, which runs along the inner side of the supinator radii longus, and sends a branch to the integuments of the thumb, and another to the dorsum of the hand, which supplies the index and middle fingers, and communicates with the dorsalis ulnaris; and a deep branch, or posterior interosseous, which supplies by superficial and deep branches the extensor muscles and terminates in a ganglion on the wrist. The median and ulnar nerves supply the flexors; the extensors are supplied by the muscular spiral nerve. This may be called the great extensor nerve.

12. *Circumflex nerve* sends branches to the

deltoid and teres minor muscles, skin, and shoulder joint. Injury of this nerve, in dislocation of the shoulder, is sometimes the cause of subsequent wasting and paralysis of the limb.

TWELVE PAIR OF DORSAL NERVES.

Posterior branches are small, and pass backwards to the muscles and integuments of the back and loins.

Anterior branches, or Intercostals. First is the largest; it contributes to the formation of the brachial plexus. Second and third run backwards and outwards, and at the angle of the ribs pass between the intercostal muscles, and running along the lower edge of each rib supply the surrounding muscles: opposite the axilla they send off the intercosto-humeral nerves, of which one, joining a branch from the brachial plexus, forms the nerve of Wrisberg, for the skin of the arm, and the terminal branches supply the muscles and skin upon the lateral and fore part of the thorax. Fourth to twelfth, inclusive, are similar to the second and third in distribution, supplying the intercostal and adjacent muscles: the two last go chiefly to the diaphragm, and the twelfth sends a branch to join the first lumbar. [All the intercostals are connected by two short branches to the ganglions of the sympathetic.]

FIVE PAIR OF LUMBAR NERVES.

They are larger than the dorsal, and like them divide into posterior and anterior branches.

Posterior branches are distributed to the lumbar muscles.

Anterior branches unite in the substance of the psoas muscle, to form the lumbar plexus.

LUMBAR PLEXUS.

It is formed by the last dorsal and four first lumbar nerves.

Branches.—1. *External inguino-cutaneous*, sends branches to the abdominal muscles, a cutaneous branch to the integuments on the outer part of the thigh, and the ilio-scrotal.

2. *Middle inguino-cutaneous*, to the abdominal muscles and skin, on the outer part of the thigh.

3. *Internal inguino-cutaneous*, or genito-crural, sends a branch to the spermatic cord, which accompanies the spermatic vessels, and a branch to the integuments and glands of the groin.

4. *Anterior crural nerve*, formed by the second, third, and fourth nerves, passes beneath Poupart's ligament, and divides into two fasciculi. The superior fasciculus sends four or five long branches, which pierce the fascia lata, and descend to the knee: some of them accompany the saphena vein; the deep fasciculus sends

external muscular branches to the vastus externus, rectus, iliacus internus muscles, and internal branches to the sartorius, vastus internus, and crureus; branches also accompany the femoral artery, near to the knee, the *internal saphenous nerve*, which joins the saphena vein at the knee, lying between the tendons of the gracilis and sartorius, goes on to the inner side of the foot, sending off numerous branches to the integuments.

5. *Obturator nerve* sends branches to the obturator internus, an anterior branch to the adductor brevis and pectineus, and a posterior branch to the gracilis, the adductor magnus and longus muscles: articular nerves are given to the hip and knee joints.

6. *Lumbo-sacral nerve* proceeds from the fourth and fifth lumbar into the pelvis, and divides into the superior gluteal and the communicating nerve: the superior gluteal is distributed to the gluteus medius, minimus, and tensor vaginæ femoris muscles: and the communicating branch joins the first sacral nerve.

FIVE PAIR OF SACRAL NERVES.

Posterior branches, very small, to muscles and integuments.

Anterior branches, very large, particularly the three superior: these five, with the communicating branch of the lumbo-sacral, form the sacral plexus.

SACRAL PLEXUS.

It is formed by the last lumbar and four first sacral nerves. Sends off internal or pelvic branches, which are named Hæmorrhoidal, Vesical, Uterine, and Vaginal. External branches.

1. *Muscular nerves*, to the pyriformis, obturator internus, gemelli, and quadratus femoris muscles.
2. *Lesser sciatic nerve*, to the gluteous maximus and integuments of the buttock.
3. *Posterior cutaneous nerve*, to the back part of the thigh and leg.
4. *Pudic nerve*, which sends an inferior branch to the muscles of the perineum and to the scrotum; and a superior branch which passes along the dorsum of the penis to its glans.
5. *Great sciatic nerve* sends off several branches to supply the semimembranosus, semitendinosus, biceps and adductor magnus muscles, also the hip joint; it divides at the upper and outer part of the popliteal space into two branches, viz. the peroneal and the popliteal nerves. The peroneal nerve first sends off the external cutaneous nerves of the leg, which communicate with the external saphenous nerve; it next gives off the musculo-cutaneous nerve, which divides into an internal and an external branch, the former being distributed to the integuments of the first and second toes, and the latter to the integuments of the third, fourth, and fifth toes; and lastly, the anterior tibial, which winds round the head of the fibula, and supplies the integuments

on the anterior part of the leg, the tibialis anticus, the extensor digitorum longus, the extensor pollicis, and the extensor digitorum brevis muscles, and terminates at the first interosseal space by communicating with the plantar nerves, having first supplied the inner interosseous muscle. The popliteal division of the great sciatic first sends off the external saphenous nerve, and branches to the gastrocnemius, soleus, plantaris, and popliteus muscle. At the lower border of the popliteus muscle it becomes the posterior tibial, which supplies the tibialis posticus, flexor longus digitorum, and flexor longus pollicis; it finally divides into the internal and external plantar nerves; the former sending branches to the plantar muscles and skin, and four digital branches to supply the 1st, 2d, 3d toes and inner side of the 4th toe, and the latter a superficial branch to the little toe and outer side of the 4th toe, and a deep branch of the plantar and interossei muscles.

SYMPATHETIC OR GANGLIONIC SYSTEM OF NERVES.

The sympathetic nerves, characterised by their reddish or grey color, and by their numerous ganglia, form a system, which communicates with all the cerebral nerves, except the three nerves of special sense [viz. the olfactory, the optic, and the auditory] and with all the spinal nerves. In the chest and abdomen large plexuses are formed in front of the vertebral column.

THE CERVICAL GANGLIONS

Are three in number:

Superior cervical ganglion extends from the first to the third cervical vertebra, and sends off 1, two superior branches, which ascend along with the carotid artery in the carotid canal to the cavernous sinus, where they communicate with the sixth or abducens nerve, and with the Vidian nerve. 2. Descending or inferior filaments which join the laryngeal and pneumo-gastric nerves, the superior cardiac nerve, and the middle cervical ganglion, if it exists; if not, they join the inferior ganglion. 3. Internal branches, which unite with the pharyngeal plexus. 4. External branches, to join the superior cervical nerves. 5. Anterior branches, which unite with the pneumo-gastric and facial nerves, and form a plexus around the carotid artery, from which branches proceed along the external carotid and its divisions.

Middle cervical ganglion, placed opposite the fifth or sixth cervical vertebra, is sometimes absent; when present it sends branches to unite with the vagus and cervical nerves, and branches to join the cardiac nerves.

Inferior cervical ganglion is situated between the transverse process of the last cervical vertebra and the neck of the first rib; it sends branches to the phrenic nerve, brachial plexus, subclavian artery, and its ramifications, and branches to the inferior cardiac nerve.

Cardiac Nerves are three in number, and are named superior, middle, and inferior. 1. Superior Cardiac Nerves arise by two or three filaments from the superior cervical ganglion, communicate with the vagus and laryngeal nerves, and with the middle and inferior cervical ganglia, pass along the coats of the arteria innominata on the right side, and between the left carotid and left subclavian on the left side of the aorta, and here communicate with the recurrent nerves, and the middle and inferior cardiac nerves, and the cardiac ganglion or plexus. 2. Middle Cardiac Nerves. That of the right side is generally the largest; on the left side it is sometimes wanting. They enter the thorax anterior to the subclavian artery, are joined by branches from the pneumo-gastric and recurrent nerves, and, passing along the arteria innominata, terminate in the cardiac ganglion and plexus. 3. Inferior Cardiac Branches. The right descends along the arteria innominata to the fore part of the arch of the aorta, and terminates in the anterior cardiac plexus; some branches pass between the aorta and pulmonary artery to the cardiac ganglion; on the left these nerves accompany the subclavian artery, and partly join the middle cardiac nerve, and partly the cardiac plexus.

Cardiac Plexus or Cardiac Ganglion is situated behind the ascending aorta, near its origin, in front of the trachea and right pulmonary artery; it consists of a plexus of nerves formed by the cardiac nerves of the opposite sides and branches of the eighth pair and recurrent. In the meshes of this plexus several small ganglions are enclosed, and to these conjointly the term cardiac ganglion is applied. Branches proceed from this plexus to the coronary and pulmonary vessels, to the aorta and vena cava, and to the substance of the heart itself.

Thoracic Ganglions are twelve on each side, sometimes only eleven, the last cervical and first dorsal being united. 1. Branches to the mediastinum, which ramify on the aorta and adjacent vessels, and communicate with the pulmonary plexus. 2. Great Splanchnic Nerve is formed by distinct roots, from the 6th, 7th, 8th, and 9th ganglions; uniting on the 9th dorsal vertebra into one cord, it enters the abdomen along with the aorta, or separated from it by a fasciculus of the diaphragm, and expands into the semilunar ganglion. 3. Lesser Splanchnic Nerve arises by two roots from the 10th and 11th ganglions; uniting on the side of the last dorsal vertebra, it enters the abdomen through the crus of the diaphragm, and ends partly in the renal plexus and partly in the semilunar ganglion.

Semilunar Ganglia are situated on the diaphragm, partly on the aorta, and on either side of the cœliac axis, and above and behind the supra-renal capsule. They are the largest ganglia of the sympathetic; several nervous filaments, on which small ganglia are placed, pass from one to the other surrounding the cœliac axis, forming a plexus, called the

Solar Plexus, which is situated behind the stomach, above the pancreas, and in front of the aorta. It gives off, 1. Branches in various directions, accompanying the blood-vessels, forming plexuses around each, and named accordingly, hepatic, splenic, and gastric, and these communicate with the eighth pair. 2. Branches descending in front of the aorta, which subdivide at the renal and mesenteric arteries, accompanying them, and forming plexuses, named, accordingly, renal, superior, and inferior mesenteric, and into each of these branches of the lumbar ganglions enter.

Renal Plexus receives the lesser splanchnic nerves; from it descends the spermatic plexus, which goes to the testicle in the male, and to the ovarium and uterus in the female.

Inferior Mesenteric Plexus sends branches which descend to the brim of the pelvis, unite with others from the lumbar ganglions, and form a plexus around the internal iliac artery and its branches, named hypogastric plexus. It is joined by numerous filaments from the lumbar and sacral ganglions of the sympathetic, and communicates with the pelvic branches of the sacral plexus.

Lumbar Ganglions are five on each side, sometimes only three or four. 1. Filaments to anterior branches of lumbar spinal nerves. 2. Filaments to assist in forming the several abdominal plexuses.

Sacral Ganglions are three or four in numer. 1. Filaments to sacral nerves. 2. Filaments to hypogastric and pelvic plexuses. 3. A small branch from the last ganglion, which passes in front of the coccyx, there forming with its fellow the

Ganglion Impar, which sends branches to the coccygeus, levator, and sphincter ani muscles.

The sympathetic nerve twines around and enters the coats of the arteries of the viscera, contained in the cranium, the neck, thorax, abdomen, and pelvis. Hence it has been called the Great Visceral nerve. It is believed to preside over the functions of secretion, nutrition, and involuntary motion.

THE THORAX AND ITS CONTENTS.

The thorax is bounded anteriorly by the sternum and cartilages of the ribs, posteriorly by the vertebræ and lesser circle of the ribs, and on either side by the shafts of the ribs and the intercostal muscles. Its upper orifice is transversely oval, and allows the exit and entrance of vessels, nerves, and muscles, to and from its cavity; its inferior orifice, or circumference, being much larger, and closed by the diaphragm.

The thorax contains the heart and lungs, and also several vessels, nerves, glands, &c., to be noticed as we proceed.

THE PLEURÆ

Are two serous membranes, one on either side, which cover the inner surface of the thorax, and are reflected upon the outer surface of the parts contained in its cavity. That portion of the pleura which lines the thorax is called the *parietal* layer, and that which lines the contained parts the *visceral layer*. Each pleura can be traced in the following manner:—From the posterior surface of the sternum it passes backwards until it meets with the anterior surface of the pericardium, along the side of which it passes to the anterior surface of the root of the lung; from this it passes upon the lung, and is reflected over the entire surface of the organ, until it arrives at the posterior surface of its root and of the pericardium, from whence it passes upon the sides of the bodies of the vertebræ, reaching as high as the transverse process of the sixth cervical vertebra on the right side, the seventh on the left, and descending to the diaphragm, the thoracic aspect of which it covers; it finally lines the ribs and intercostal muscles, until it arrives at the portion which was opened, and which corresponds to the posterior aspect of the sternum.

Ligamentum latum pulmonis (one on either side) is merely a triangular fold of pleura, formed by the reflection of the membrane from the lower edge of the root of the lung upon the vessels from the heart.

ANTERIOR MEDIASTINUM.

A triangular cavity formed by tearing through the cellular tissue, which connects the right and left pleura behind the sternum; the base is formed by the sternum, the sides by the separated pleuræ, and the apex corresponds to the anterior surface of the pericardium, where the pleuræ separate to enclose this bag. Thus formed, it contains the origins of the sterno-hyoid and sterno-thyroid muscles, the remains of the thymus gland with its vessels, lymphatic glands and absorbents, the triangularis, sterni muscles, and loose cellular tissue.

MIDDLE MEDIASTINUM

Is of oval shape, and is formed by the reflection of the pleuræ upon the sides of the pericardium; it consequently contains this bag and its contents, viz. the heart, with its vessels; also the phrenic nerves.

POSTERIOR MEDIASTINUM

Is formed by the reflection of the pleuræ upon the sides of the bodies of the vertebræ; it is of triangular form, the apex anterior corresponding to the posterior surface of the pericardium, the sides formed by the pleuræ, and the base represented by the anterior surfaces of the bodies of the vertebræ; it extends from the third to the tenth dorsal vertebra, and contains the following parts:—the bifurcation of the trachea, the œsophagus and pneumogastric nerves, the thoracic duct, the vena azygos, the thoracic aorta, lymphatic glands, absorbents, bronchial and œsophageal arteries, and loose cellular tissue.

THE LUNGS

Are two soft, spongy, vascular bodies, one contained on each side of the cavity of the chest. Each lung resembles a cone, with that side corresponding to the median line truncated; the base, concave, corresponds to the diaphragm; the obtuse rounded apex rises in the neck, a little above the level of the first rib; the external convex surface corresponds to the internal concave surface of the thoracic parietes, and the flat or truncated surface corresponds to the mediastina. The posterior edge of the lung is thick and rounded, whilst the anterior is thin and irregular. Each lung is distinguished into lobes, which are separated from each other by fissures; a little above the centre of each is the *root* formed by the pulmonary vessels and bronchial tube, connected to each other by cellular tissue, and invested by the pleura. The bronchial tube is situated posterior and superior to the pulmonary vessels; the two pulmonary veins are placed anterior and inferior to the artery and bronchus, and the pulmonary artery is placed between the bronchus and the pulmonary veins, but behind the pulmonary veins and before the bronchus. On the left side, the bronchus, more oblique than its fellow, descends near the root of the lung so as to lie between the artery and vein. The *root* of each lung has anterior to it the phrenic nerve and filaments of the pneumo-gastric nerve, posterior to it the pulmonic plexus. The root of the right lung has the vena azygos arching over it.

The right and left lungs differ from each other in some important particulars: the right lung is broader and shorter than the left, and consists of three lobes, separated by two fissures; the right also ascends higher in the neck, and the anterior edge of the left presents a notch where it corresponds to the apex of the heart.

The intimate structure of the lungs consists of the ultimate ramifications of the bronchial tubes, which are the continuations of the trachea, and the branches of the pulmonary artery and veins; the larger tubes at the root of the lung receive bronchial arteries for their nutrition.

TRACHEA AND ITS RAMIFICATIONS.

The wind-pipe, or trachea, is a cylindrical tube, extending from the cricoid cartilage of the larynx to the level of the third dorsal vertebra. It consists of from seventeen to twenty fibro-cartilaginous rings, truncated behind, and connected to each other by an elastic membrane; about the posterior fourth of each ring is deficient, and its place is supplied by fibrous membrane and unstriped or involuntary muscular fibre.

Opposite the third dorsal vertebra the trachea

divides into the right and left bronchial tubes; the right bronchus, larger than the left, runs transversely into the root of the lung and divides into three branches; the left bronchus passes under the arch of the aorta, to the root of the left lung, and divides into two branches.

The bronchia consists of cartilaginous rings, but as these tubes advance into the substance of the lung they diminish in size and firmness, until their place is supplied by fibrous tissue, involuntary muscular fibres, which tissue also disappears, and at length nothing remains but the mucous membrane, which terminates in the air-cells, upon which ramify the ultimate ramifications of the pulmonary artery and the commencing radicles of the pulmonary veins.

The ramifications of the pulmonary artery communicate with those of the pulmonary veins, beneath the mucous membrane of the air-cells, and are enveloped in fine cellular tissue; and except this cellular tissue the lung has no proper parenchyma, its structure being entirely vascular. The roots of the lungs are supplied with blood by the *bronchial arteries*, derived from the thoracic aorta; these vessels run along the bronchial tubes, subdivided as they proceed, and form a minute net-work on the attached surface of the bronchial mucous membrane; the blood they convey to the lungs is returned to the vena azygos, or superior intercostal. The nerves distributed to the lungs are derived from the eighth pair, and from the sympathetic.

HEART AND PERICARDIUM.

The pericardium consists of two layers, an outer or proper fibrous layer, and an inner, or serous layer. It is of conical form, the base below connected to the central division of the cordiform tendon of the diaphragm, the apex above corresponding to the great vessels at the base of the heart, along the outer coats of which the fibrous layer is gradually lost; it is connected laterally to the pleura and to the pulmonary vessels; the phrenic nerves, one on either side, run in close contact with it to the diaphragm. Upon laying open the cavity of the pericardium, the serous layer is exposed, and, like all serous membranes, it consists of two portions—a parietal layer, which lines the inner surface of the fibrous pericardium, and a visceral layer, which lines the outer surface of the heart and great vessels. When the pericardium is fully opened, we bring into view, covered by the serous membrane, the right auricle, the venæ cavæ, the left auricular appendix, the right ventricle, the tip of the left ventricle (which forms the apex of the heart), the aorta, the pulmonary artery, and the anterior branches of the coronary vessels, with the ramifications of the cardiac nerves. Upon turning up the heart its posterior surface will be brought into view, presenting the left auricle (proceeding to which, on either side, are the pulmonary veins) and the left ventricle.

The serous membrane may be traced in the

following manner:—After having lined the fibrous pericardium it is reflected on the superior cava and aorta, and the pulmonary artery, as these vessels are passing through the fibrous membrane, ascending highest, however, upon the aorta; inferiorly it is partly reflected around the inferior cava, as this vessel pierces the fibrous pericardium to enter the right auricle, and laterally it is reflected upon the pulmonary veins as these vessels pierce the fibrous pericardium to enter the left auricle; from these different points it reaches the surface of the heart, which it completely covers.

THE HEART, of conical shape, is situated obliquely between the lungs, its base being superior, posterior, and to the right side, its apex pointing towards the cartilage of the sixth rib of the left side. The axis of the heart is obliquely from right to left, and from behind forwards. It is retained in its situation by the great vessels and the reflections of the serous membrane.

The heart consists of four cavities, two auricles and two ventricles; the auricles are separated from each other by a partition, called septum auricularum, the ventricles by the septum ventriculorum. We shall examine these cavities in the order of the circulation.

THE RIGHT AURICLE is placed between the two venæ cavæ, the blood conveyed by which it receives and transmits to the right ventricle; the small loose portion is called the *auricular appendix*, and the portion between the cavæ the *sinus of the auricle*. Upon laying open this auricle, by a perpendicular cut from the superior cava to within a few lines of the entrance of the inferior cava, and by a second cut from the centre of this at right angles towards the auricular appendix, the following parts present themselves.

Tuberculum Loweri, an eminence upon the under surface of the sinus of the auricle, and placed between the orifices of entrance of the venæ cavæ.

Septum auricularum, a membrano-muscular partition separating this auricle from the left.

Fossa ovalis, an oval depression in the septum, which marks the situation of the foramen ovale, or the oval communication which existed between both auricles in the fœtus. The edges of this fossa present a thickened margin.

Eustachian valve, a semilunar fold of the lining membrane, the anterior aspect of the inferior vena cava.

Musculi pectinati, the muscular bands situated in the auricular appendix.

Openings of the Venæ Cavæ. The superior cava opens at the upper and back part of the auricle, its direction being downwards, forwards, and inwards; the inferior cava opens beneath, in a direction upwards, backwards, and inwards.

Opening of the Coronary Vein is situated between the Eustachian valve and the right ventricle, and is guarded by a semilunar valve (the valve of the coronary vein).

Opening of the Auricular Appendix is small

and circular, and exists where this portion of the auricle joins the sinus.

Foramina Thesebii, small orifices on different parts of the auricle, supposed to be the openings of veins.

Right Auriculo-ventricular opening, the large opening by which the auricle communicates with the ventricle, the boundaries of each cavity being marked by a white line.

RIGHT VENTRICLE, of conical form, is joined by its basis to the right auricle, its apex being above the apex of the heart, in consequence of the apex of this organ being formed by the left ventricle.

Septum ventriculorum, a thick muscular partition which separates one ventricle from the other.

Carneæ columnæ, the muscular projections in the interior of the ventricle, which give it its irregular appearance.

Chordæ tendineæ, the delicate but strong tendinous cords which are connected by one extremity to the carneæ columnæ, and by the other to the tricuspid valve.

Tricuspid Valve, three triangular duplicatures of the lining membrane of the heart, strengthened by the chordæ tendineæ which pass from their apices to their bases. These triangular valves are attached by their bases to the right auriculo-ventricular opening, and by their apices to the chordæ tendineæ, and prevent the regurgitation of the blood from the ventricle into the auricle, by closing the opening of communication between both cavities, when the ventricle contracts.

Orifice of the Pulmonary Artery is situated at the left extremity of the base of the ventricle, close to the right auriculo-ventricular opening, which is situated inferior, posterior, and to its right, and separated from the mouth of the artery by the largest triangular division of the tricuspid valve.

Pars planum is a name given to that smooth portion of the ventricle which leads to the mouth of the pulmonary artery.

Pulmonic Semilunar Valves extend from the line of junction of the pulmonary artery and right ventricle into the cavity of the former. These three valves are duplicatures of the lining membrane, and are attached by their convex edges to the roots of the pulmonary artery; their free concave margins presenting in their centre a small tubercle, called *corpus sesamoideum*, or *corpus arantii.* These valves, when the blood passes from this ventricle into the pulmonary artery, are thrown down, and thus, by closing the opening, prevent a reflux of the blood back again into the ventricle.

LEFT AURICLE is placed at the upper and back part of the heart and is of quadrilateral form. It is smaller than the right auricle, and receives, at its four angles, the openings of the pulmonary veins; its parietes are much thicker than the right; its auricular appendix smaller, but the musculi pectinati are the stronger: with the exception of the auricular appendix its inner surface is smooth, and it communicates with its corresponding ventricle by the left auriculo-ventricular opening, which is situated inferior to the opening of the auricular appendix, and is marked by a whitish line. The septum auricularum presents towards this auricle a smooth aspect, from circumstances already explained.

LEFT VENTRICLE is longer, stronger, and smaller than the right; from its greater length it forms the apex of the heart; from its greater strength it is of power sufficient to propel the arterial blood through the aorta and its ramifications. We consequently find the carneæ columnæ, the chordæ tendineæ, the bicuspid, or mitral valve, the parietes of the cavity, the aorta, which arises from it, and the semilunar valves with their corpora sesamoides which guard the opening of this vessel, much stronger than in the right division of the heart.

The left auriculo-ventricular opening, and the mouth of the aorta, are situated at the base of this cavity, and are close to each other, the aortic opening being anterior, and both being separated, as in the right ventricle, by the larger division of the bicuspid valve. With the exception of the peculiarities just mentioned, the left auricle and ventricle are, in their anatomical configuration, similar to the right, and the several valves, muscular eminences, tendons, cords, &c., serve similar purposes: the septum ventriculorum belongs chiefly to the left ventricle.

ARTERIES.

THE AORTA

Is divided into three portions—viz., the arch of the aorta, the thoracic aorta, and the abdominal aorta.

THE ARCH OF THE AORTA

Sends off five branches.

A. *Arteria coronaria dextra*, which sends a branch to the right auricle, a branch to the anterior part of the right ventricle, and a branch to the posterior part of the right ventricle.

B. *Arteria coronaria sinistra*, which sends a branch to the left auricle, and a branch to the left ventricle.

C. *Arteria innominata*, which divides into the right carotid and right subclavian arteries.

D. *Left carotid artery.*

E. *Left subclavian artery.*

THE COMMON CAROTIDS

Divide opposite the upper corner of the thyroid cartilage into two branches.

A. *External carotid artery.*

B. *Internal carotid artery.*

Sends off ten branches, viz.:—

A. *Superior thyroid*, which sends off—1st, a hyoidean branch; 2d, a superficial branch; 3d, a laryngeal branch; and 4th, a thyroidean branch.

B. *Lingual*, which sends off—1st, a hyoidean branch; 2d, the dorsalis linguæ artery; 3d, the sublingual artery; and 4th, the ranine artery.

C. *Facial*, which sends off—1st, the inferior palatine; 2d, the tonsillar; 3d, the glandular; 4th, the submental; 5th, the inferior labial; 6th, the inferior coronary; 7th, masseteric; 8th, superior coronary; 9th, the lateral nasal; and 10th, the angular, which anastomoses with the ophthalmic.

D. *Muscular*.

E. *Occipital*, which gives off the princeps cervicis, the inferior meningeal, and the terminal branches, of which one enters the mastoid foramen; the others ramify in the scalp.

F. *Posterior auris*, which gives off the muscular, the glandular, and the stylo-mastoid.

G. *Ascending pharyngeal*, which gives off the pharyngeal branches and the meningeal branches.

H. *Transverse facial*.

I. *Superficial temporal*, which gives off—1st, the anterior auris; 2d, the capsular branches;

3d, the middle temporal; 4th, the posterior temporal; and 5th, the anterior temporal.

K. *Internal maxillary*, which gives off—1st, the tympanic, which enters the fissura Glasseri; 2d, the meningea media, which enters the foramen spinosum; 3d, the inferior maxillary or dental, which enters the dental canal; 4th, the deep temporal; 5th, the pterygoid; 6th, the masseteric; 7th, the buccal; 8th, the alveolar; 9th, the infra-orbital; 10th, the descending palatine; 11th, the vidian; 12th, the pterygo-palatine; 13th, the spheno-palatine.

THE INTERNAL CAROTID

Supplies no vessels in the neck, but in its passage through the petrous portion of the temporal bone, gives off—

A. *The tympanic.*

B. *The arteræ receptaculi*, or vessels to the cavernous sinus.

C. *The anterior meningeal.*

Opposite the anterior clinoid process it divides into—

A. *Ophthalmic artery*, which sends off—1st, the lachrymal; 2d, the centralis retinæ; 3d, the supra-orbital; 4th, the ciliary branches; 5th, the muscular; 6th, the posterior ethmoidal; 7th, the anterior ethmoidal; 8th, the palpebral; 9th, the nasal; and 10th, the frontal.

B. *Anterior cerebral*, which sends off—1st,

the anterior communicans; 2d, the arteria corporis callosi.

C. *Arteria media cerebri.*
D. *Posterior communicans.*

The anterior cerebral arteries are united by the anterior communicans, the posterior communicans passes back to join the posterior cerebral (a branch of the basilar), and in this manner is formed an arterial circle, called the circle of Willis.

THE SUBCLAVIAN ARTERY

Is divided for surgical purposes into three portions—1st, on the *right* side it extends from the arteria innominata to the inside of the scalenus anticus. 2d, the second portion lies *under* the scalenus anticus. 3d, the third portion extends from the scalenus anticus to the lower border of the first rib. On the *left* side, the left subclavian extends from the arch of the aorta to the inside of the scalenus anticus muscle. 2d, the second portion of the left subclavian, like that of the right side, lies *under* the scalenus anticus. 3d, the third portion of the left subclavian, also like that of the right, extends from the scalenus anticus to the lower border of the first rib. The *third* portion of the subclavian arteries is the one usually ligated.

A. *Vertebral*, which gives off—1st, the arteriæ medullæ spinalis transversæ; 2d, the meningeal; 3d, the inferior cerebellar; 4th, anterior and posterior spinal. The basilar artery, formed by the union of the two vertebrals, gives off transverse branches to the pons, a small branch enters the meatus auditorius internus; and terminates by dividing into four large vessels—the posterior cerebral and superior cerebellar of either side.

B. *Internal mammary*, which gives off—1st, the anterior intercostal; 2d, the mediastinal; 3d, the comes nervi phrenici; 4th, the musculo-phrenic; and 5th, the superior epigastric.

C. *Thyroid axis*, which divides into—1st, inferior thyroid, which gives off the cervicalis ascendens; 2d, supra-scapular, which gives off a supra-acromial branch, and then passes into the supra-spinous fossa of the scapula, over the notch; 3d, posterior scapular, which gives off a superficial cervical branch, and then terminates in the muscles of the scapula.

D. *Cervicalis profunda*, anastomoses with princeps cervicis from the occipital.

E. *Superior intercostal*, supplies the two or three first intercostal spaces.

THE AXILLARY ARTERY

Extends from first rib to the lower border of the tendon of the latissimus dorsi; it sends off seven branches.

A. *Acromial thoracic.*
B. *Thoracica suprema.*
C. *Thoracica alaris.*
D. *External Mammary*, or long thoracic.

E. *Subscapular*, which sends off an anterior and a posterior branch.
F. *Posterior circumflex.*
G. *Anterior circumflex.*

THE BRACHIAL ARTERY

Extends from the lower border of the tendon of the latissimus dorsi to a finger's breadth below the bend of the elbow, and sends off four branches.

A. *Profunda superior*, which sends off—1st, an ascending branch; and 2d, the musculo-spiral branch. It accompanies the musculo-spiral nerve.
B. *Nutritia humeri.*
C. *Profunda inferior*, which accompanies the ulnar nerve.
D. *Anastomotica magna.*

THE RADIAL ARTERY

Sends off ten branches.

A. *Recurrent radial.* It anastomoses with the profunda superior.
B. *Muscular.*
C. *Superficialis volæ*, anastomoses with the ulnar to form the superficial palmar arch.
D. *Anterior carpi radialis.*
E. *Dorsalis carpi radialis.*
F. *Dorsalis pollicis.*
G. *Dorsalis indicis.*
H. *Princeps pollicis.*

I. *Radialis indicis.*
K. *Palmaris profunda.*

THE ULNAR ARTERY

sends off eight branches.

A. *Anterior recurrent.*
B. *Posterior recurrent.*
C. *Interosseous,* which sends off—1st, the anterior recurrent; 2d, the anterior interosseous; 3d, the posterior interosseous; 4th, the posterior recurrent; and 5th, the posterior descending branch.
D. *Muscular.*
E. *Carpi ulnaris anterior.*
F. *Ulnaris posterior.*
G. *Arteria communicans.*
H. *Palmaris superficialis,* which anastomoses with the superficialis volæ to form the superficial palmar arch. A long branch comes off from the ulnar or interosseous, called the comes nervi mediani; it is of uncertain size.

PALMAR ARCHES.

The deep palmar arch is formed by the palmaris profunda of the radial, uniting with the arteria communicans from the ulnar; it sends off five small branches to supply the interossei muscles.

THE SUPERFICIAL PALMAR ARCH

is formed by the arteria palmaris of the ulnar, uniting with the superficialis volæ from the radial. It sends off four branches.

A. *Branches to ulnar edge of little finger.*
B. *Branch to cleft between little and ring fingers.*
C. *Branch to cleft between ring and middle fingers.*
D. *Branch to cleft between middle and index fingers.*

THE THORACIC AORTA

sends off five sets of branches.
   A. *Pericardiac.*
   B. *Mediastinal.*
   C. *Bronchial.*
   D. *Œsophageal.*
   E. *Intercostals:* each divides into—1st, the posterior branches; and 2d, the anterior branches.

THE AORTA ABDOMINALIS

sends off the following branches.
   A. *The two phrenic arteries.*
   B. *The cœliac axis.* From this axis arise—1st, the superior gastric artery, which divides into a superior and an inferior branch; 2d, the hepatic artery, which gives off the superior pyloric artery, the gastro-duodenal artery, which divides into the arteria pancreatica duodenalis, and the arteria gastro-epiploica dextra. The hepatic artery then divides into the left hepatic and right hepatic arteries, from the last of which proceeds a small branch to the gall-bladder, called arteria cystica; 3d, the splenic artery, which sends off the pancreaticæ parvæ, the pancreatica

magna, the vasa brevia, splenic branches and the gastro-epiploica sinistra.

C. *The superior mesenteric artery*, which gives off—1st, the colica dextra, which divides into a superior and an inferior branch; 2d, the colica media, which divides into a right and left branch; 3d, the ileo-colica, which divides into a superior branch, a middle branch, and an inferior branch; and lastly, the mesenteric branches, from fifteen to twenty in number.

D. *Two capsular*, to the renal capsules.

E. *Two renal*, to the kidneys.

F. *Two spermatic*, to the testicles.

G. *Inferior mesenteric*, which sends off—1st, the colica sinistra, which divides into an ascending branch and a descending branch; 2d, the sigmoid artery; and 3d, the superior hemorrhoidal artery.

H. *Ureteric arteries.*

I. *Lumbar arteries.*

K. *Sacra media.*

THE COMMON ILIAC ARTERIES.

extend from the bifurcation of the aorta to opposite the sacro-iliac symphasis; the right is usually longer than the left: they extend into two branches, viz.

The internal, and
External iliac arteries.

THE INTERNAL ILIAC ARTERY

sends off eleven branches in the female.

A. *Arteria ilio-lumbalis*, which sends off—1st, ascending branches; 2d, external branches; and 3d, descending branches.

B. *Lateral sacral.*

C. *Middle hæmorrhoidal.*

D. *Vesical.*

E. *Umbilical.*

F. *Uterine.*

G. *Vaginal.*

H. *Obturator*, which sends off—1st, the branches within the pelvis; and 2d, branches without the pelvis.

I. *Gluteal*, which sends off—1st, a superficial branch; and 2d, a deep branch.

K. *Ischiatic*, which sends off—1st, the coccygæal branch; 2d, the arteria comes nervi ischiatici; and 3d, the muscular branches.

L. *Pudic*, which sends off—1st, the external hæmorrhoidal arteries; 2d, the perinæal; 3d, the transversalis perinæi; 4th, the arteria corporis bulbosi, which gives a branch to Cowper's gland, and a branch to the corpus spongiosum; 5th, the arteria corporis cavernosi penis; and 6th, arteria dorsalis penis.

THE EXTERNAL ILIAC,

sends off two branches.

A. *Epigastric*, which sends off—1st, a spermatic branch, and, 2d, the muscular arteries. The epigastric anastomoses with the internal mammary.

B. *Circumflexa-ilii*, anastomoses with the ilio-lumbar.

THE FEMORAL ARTERY

extends from Poupart's ligament to an opening in the tendon of the adductor magnus, where it becomes "popliteal." At Poupart's ligament, the femoral vein is on the inside of the femoral artery, the anterior crural nerve is on the outside of the artery. In the upper third of the thigh, the femoral artery lies in a triangular (Scarpa's space) space, bounded on the outer side by the sartorius, inside by the adductor longus brevis and the pectineus, and superiorly by Poupart's ligament. In the middle of the thigh the artery is overlapped by the sartorius. In ligaturing the femoral artery here, an incision is to be made on the inside of the sartorius. It sends off the following branches:

A. *Superficial epigastric.*
B. *Pudendæ superficialis.*
C. *Circumflexo ilii superficiales.*
D. *Profunda femoris*, which sends off—1st, the circumflexa externa, from which arise the ascending branches, the circumflex branches, and descending branches; 2d, the circumflexa interna, from which arise arterial branches to the muscles of the hip-joint, and a branch to the interior of the hip-joint; 3d, the arteria perforans prima; 4th, the perforans secunda; 5th, the perforans tertia.
E. *Anastomotica magna.*

THE POPLITEAL ARTERY

lies upon the ligamentum posticum Winslowii, and upon the popliteus muscle: and divides into anterior and posterior tibial; it sends off seven branches.

A. *Superior Muscular.*
B. *Articularis superior externa,* which sends off a superficial branch, and a deep branch.
C. *Articularis superior interna,* which sends off a superficial branch, and a deep branch.
D. *Azygos branch,* which perforates the ligamentum posticum Winslowii.
E. *Articularis inferior externa.*
F. *Articularis inferior interna.*
G. *Inferior muscular.*

THE TIBIALIS POSTICA

lies upon the tibialis posticus, and flexor longus digitorum, and is bound down by the deep fascia of the leg. It is accompanied by two veins (venae comites) and by the posterior tibial nerve —sends off—

A. *Muscular.*
B. *Peroneal,* which sends off—1st, the nutritia fibulæ; 2d, the muscular; 3d, the anterior peroneal; and 4th, the posterior peroneal.
C. *Nutritia tibiæ.*
D. *Internal plantar.*
E. *External plantar.*

THE TIBIALIS ANTICA

passes through the interosseous space, and sends off the following branches.
  A. *Muscular.*
  B. *Recurrent.*
  C. *Internal malleolar.*
  D. *External malleolar.*
  E. *Tarsal branch.*
  F. *Metatarsal branches.*
  G. *Arteria pollicis.*
  H. *Arteria communicans.*

THE PLANTAR ARCH

is formed by the external plantar artery uniting with the arteria communicans of the anterior tibial; it sends off two sets of branches:—
  A. *Arteriæ perforantes.*
  B. *Arteriæ digitales.*

THE PULMONARY ARTERY

divides into two branches:—
  A. *Right pulmonary.*
  B. *Left pulmonary.*

VEINS.

The veins are not so uniform in their distribution as the arteries. Besides the numerous superficial veins which ramify on the skin, one or two are found to accompany each artery. In the extremities there are generally two to each artery, and in these situations they receive the name of *venæ comites*. When, however, an artery is of great size, as the femoral or the axillary, it is accompanied by but one vein, which receives the same name as the artery.

VEINS OF THE HEAD AND NECK.

The veins which accompany the branches of the internal maxillary artery, form the internal maxillary vein.

External jugular vein, formed by the junction of the internal maxillary and one temporal vein, descends obliquely backwards, and joins the subclavian vein; after crossing the sternomastoid muscle it receives the supra and posterior scapula veins.

Internal jugular vein commences at the termination of the lateral sinus, descends along the outer side of the common carotid artery, and joins the subclavian vein at the sternal extremity of the clavicle. It receives the superior thyroid, lingual, facial, occipital, and ascending pharyngeal veins.

VEINS OF THE SUPERIOR EXTREMITY.

The Cephalic vein is formed by the union of several veins from the back of the hand. It ascends along the radial side of the extremity, and, passing along the interval between the pectoralis major and deltoid muscles, terminates in the *axillary vein.*

The Basilic vein commences by a small vein from the little finger (*vena salvatella*), ascends along the inner side of the extremity, and terminates in the axillary vein, or joins one of the *venæ comites* which accompany the brachial artery.

The Median vein commences at the fore part of the wrist and hand, ascends along the anterior aspect of the fore-arm, and at the elbow terminates by dividing into two branches. One of these joins the basilic vein, and is named the *median basilic;* the other joins the cephalic vein, and is named *median cephalic.*

The Axillary vein, formed by the union of the veins just described, and by the *brachial venæ comites,* ascends in front of the axillary artery, receiving the *thoracic veins,* and passes beneath the clavicle, where it terminates in the subclavian vein.

The Subclavian vein passes inwards over the anterior scalenus muscle, receives numerous veins from the neck and shoulder, also the *external jugular* and *vertebral veins,* and joins the internal jugular vein behind the sternal extremity of the clavicle.

Vena innominata, formed by the union of the internal jugular and subclavian veins, on the right side is very short, and descends into the thorax; the left vena innominata, which is much longer, enters the thorax in a transverse direction in front of the trachea to join the right vein, and in its course receives the *thyroid veins,* and veins from the anterior mediastinum. By the union of the venæ innominatæ is formed the

Vena cava superior, which descends in front of the right pulmonary vessels, enters the pericardium, and opens into the right auricle.

VENA AZYGOS,

commences on the first lumbar vertebra by one or two small veins from the renal or from the inferior vena cava, gains the posterior mediastinum by passing through the aortic opening of diaphragm, ascends along the right side of the bodies of the dorsal vertebræ, curves forwards over the root of the right lung, and terminates at the posterior aspect of the superior vena cava, as this vessel is entering the pericardium. In its course it receives the lower intercostal veins of each side, œsophageal veins, the azygos minor, right bronchial, and sometimes the right superior intercostal veins. It has no valves.

VEINS OF THE INFERIOR EXTREMITY.

Internal saphena vein, commences at the inner part of the foot, ascends along the inner side of the leg and knee, behind the inner condyle, becomes more anterior upon the thigh, and reaching to within about two inches of Poupart's ligament, passes through the saphenic opening of the fascia lata, and joins the femoral vein.

External saphena vein, commences at the outer part of the foot, ascends along the back part of the leg and ham, and joins the popliteal vein.

Both these veins are superficial, and in their course receive several veins from the integuments.

Deep veins of the leg, two veins (venæ comites) accompany each artery in the leg, and terminate in the following:—

Popliteal vein: this vessel accompanies the artery of the same name, and having received the external saphena vein, and the veins of the knee, terminates in the femoral vein.

Femoral vein accompanies the femoral artery, and having received the profunda vein, the internal saphena vein, and a few muscular veins, passes beneath the crural arch, and ends in the external iliac vein.

External iliac vein accompanies the external iliac artery. This vein is below, and to the inside of the artery.

Internal iliac vein is formed by the union of the veins which accompany the branches of the internal iliac artery, and joins the external iliac vein at the sacro-iliac symphysis.

Common iliac vein, formed by the union of the internal and external iliac veins, joins its fellow at the right side of the body of the fourth lumbar vertebra to form the inferior vena cava.

Inferior vena cava ascends along the right side of the bodies of the lumbar vertebræ, behind the liver, and passes through the quadrilateral opening in the tendon of the diaphragm, and opens into the right auricle of the heart at its lower and back part. It receives the middle sacral, one, and sometimes both spermatic veins, the emulgent and capsular veins, the venæ cavæ hepaticæ, and the phrenic veins.

VENA PORTÆ

commences on the back of the rectum by one of the *hæmorrhoidal* veins, ascends towards the meso-colon, and unites with the *inferior mesenteric vein;* this trunk next unites with the *superior mesenteric vein*, and behind the pancreas it unites with the *great splenic vein*, and receives veins from the stomach, duodenum, and pancreas. The vena portæ thus formed ascends to the right side, inclosed in the capsule of Glisson, and reaches the transverse fissure of the liver, where it divides into two branches: these enter the liver, ramify through its substance like an artery, and its blood is returned to the inferior vena cava by the venæ cavæ hepaticæ, which are three or four in number, and open into the cava as this vessel is entering its opening in the diaphragm. The vena portæ represents a tree, the roots of which are in the stomach, intestines, pancreas and spleen; they gradually converge to form a large trunk, which passes up to the transverse fissure of the liver, and divides into branches which subdivide like the branches of a tree, and terminate in the minute lobules of the liver.

THE DIGESTIVE APPARATUS.

THE MOUTH.

This cavity is bounded superiorly by the hard and soft palate, inferiorly by the tongue and the reflections of mucous membrane from it to the gums, and laterally by the cheeks. Its anterior opening, which forms the commencement of the digestive canal, corresponds to the lips; and posteriorly it communicates with the pharynx through the opening called isthmus faucium, the boundaries of which are the soft palate and uvula superiorly, the tongue inferiorly, and the pillars of the palate and tonsils laterally. Within the mouth we observe the opening of the three salivary glands and follicles, and the teeth of either side. The parotid gland opens by a single orifice opposite the second superior molar tooth; the sub-maxillary gland of either side, by a single orifice at the anterior part of the tongue, by the side of the reflection of mucous membrane, called frenum linguæ; and the third salivary gland or sublingual by several small orifices (eight or ten) on either side of the frenum linguæ.

THE TEETH.

The number of teeth in the adult is thirty-two, sixteen in each jaw; and to distinguish them from the teeth of the child they are called *permanent*. They are divided into eight *inci-*

sors, four *canine,* eight *bicuspids,* and twelve *molars.* The last molar does not appear until late in life, and hence has been called the *wisdom* tooth. It is, however, sometimes wanting.

In the child the number of teeth is only twenty, and are called milk, deciduous, or temporary teeth. They are divided into eight *incisors,* four *canine,* and eight *molars.*

THE PHARYNX

is a musculo-membranous bag of conical shape, extending by its base from the posterior part of the mylo-hyoid ridge and base of the skull to the posterior aspect of the cricoid cartilage, where it terminates in the œsophagus. It is connected by its posterior wall to the vertebræ by loose cellular tissue, and interiorly it corresponds to the mouth and larynx.

On laying open the cavity of the pharynx by a perpendicular incision along its posterior median line, the internal mucous lining of the bag will be exposed, and the following openings, viz., superiorly, one on either side of the mesial line, the openings of the posterior nares; more externally, one on either side of the openings of the Eustachian tubes; inferior to these is the opening of the mouth into the pharynx, or the isthmus faucium; posterior and inferior to the tongue is the superior opening of the larynx, and, lastly, the opening of the pharynx into the œsophagus.

THE ŒSOPHAGUS

extends from the termination of the pharynx to the stomach; it is placed above, between the vertebræ and the trachea, inclines at the inferior part of the tube to the left side, and passes behind its left bifurcation to reach the posterior mediastinum. In the mediastinum it descends forwards, above the thoracic aorta, passes through the oval muscular aperture of the diaphragm, to terminate in the stomach. The mucous membrane of the mouth, pharynx, and œsophagus, is covered by epithelium.

THE STOMACH.

A large pyriform musculo-membranous bag, situated in the epigastric and left hypochondriac regions, communicating at one extremity with the œsophagus, and at the other with the duodenum.

Connexions.—Its large extremity, or *fundus*, to the spleen by the gastro-splenic omentum; its upper concave, small edge, to the liver, by the gastrohepatic omentum; and its lower, convex large edge, to the colon by the gastro-colic omentum. Its œsophageal or cardiac orifice, situated between the fundus and lesser curve, connects it to the diaphragm, and its pyloric orifice to the duodenum. The superior-anterior surface of the stomach looks towards the diaphragm, ribs, and left lobe of the liver; the posterior-inferior surface towards the meso-colon.

The stomach is composed of three tunics, connected by cellular tissue, an external peritoneal coat, an internal mucous coat, and between both the muscular coat. The muscular fibres of the stomach observe three directions; the longitudinal are seen along the edges or curves, the oblique on the fundus, and the circular are well developed at the centre of the organ, and at its pyloric orifice. The mucous coat, smooth, and of light pink color, is thrown into folds (*plicæ*), which intersect each other, inclosing irregular quadrilateral spaces. Numerous mucous glands are found along the curves and at the pylorus: in the fundus small glands exist, which secrete the gastric juice. At the pyloric orifice the mucous membrane is thrown into a circular fold, which forms an imperfect valve between the stomach and duodenum; and at the œsophageal opening the epithelial lining is observed to terminate in a fringed edge.

SMALL INTESTINES

are divided into duodenum, jejunum, and ileum.

Duodenum, the shortest portion of the small intestines, forms a curve in the concavity of which is situated the head of the pancreas; it is divided into a superior transverse portion, a middle perpendicular portion, and an inferior transverse portion. The transverse portion mounts upwards and to the right to the under surface of the liver and neck of the gall-bladder; the descending extends as low as the body of the third lumbar vertebra; the transverse portion crosses to the left side of the body of the second lumbar vertebra, and then opens into the jejunum. The superior part is covered by peritoneum on both its surfaces, and on this account is more movable than the perpendicular or inferior portions.

Besides the numerous mucous glands which open on its interior surface, the pancreatic duct and the common biliary duct enter at its perpendicular division, either by a common aperture, or in close proximity. These ducts generally open about from three to four inches below the pyloric orifice of the stomach. This intestine differs not only in these particulars from the rest of the small intestines, but also in being much larger, more dilatable, more fixed to its position, in having a greater number of valvulæ conniventes (or circular folds of the mucous membrane), and in its muscular fibres being much stronger. It also contains Brunner's glands, which are small conglomerate glands near the pylorus.

The duodenum has also been called *ventriculus succenturiatus*.

Jejunum and Ileum form the longest part of the intestinal tube, being in general from 24 to 30 feet in length; the upper two-fifths are given to the jejunum, and the remainder to the ileum, but there is no anatomical foundation for this arbitrary boundary, as the intestines run into each other insensibly, and from the duodenum the remainder of the small intestinal tube gradually diminishes in thickness, has fewer valvulæ conniventes, and exhibits less vascularity; so much so, that the termination of the ileum is much thinner and paler than the upper part of the jejunum, and it is in these situations only that the differences between both are marked and striking. The mucous membrane of the small intestine is studded with glands of two orders, viz. the *glandulæ solitariæ*, and the *glandulæ agminatæ*. The solitariæ are disseminated like granules over the mucous membrane, and the agminatæ are placed in oval clusters. These last sets of glands are termed Peyer's glands. The fibres of the muscular coat of the small intestines take a circular and longitudinal direction, the latter being placed externally.

Peyer's plates or glands, situated in the lower part of the ileum; they are often ulcerated in typhus fever, and sometimes in tubercular disease of the lungs; they do not always exist.

LARGE INTESTINE,

divided into the cœcum, colon, and rectum, forms about one-fifth of the intestinal canal. It differs from the small intestine in its great size, in being sacculated, in having small processes along its entire course, called *appendices epiploicæ* in presenting three well defined longitudinal bands, in being thinner, and in having no valvulæ conniventes. It is composed of an external serous coat, an internal mucous coat, and between both, a muscular coat. The fibres of the muscular coat are longitudinal and circular; the former are collected into three bands, which being shorter than the intestine throw it into sacculi: the latter resemble the circular fibres of the small intestine.

Cæcum or *caput coli*, placed in the right iliac fossa and connected to the iliacus and psoas muscles, is fixed in its situation by the peritoneum, which only covers it anteriorly and somewhat laterally; it receives at its inner side the ileum, which terminates in its cavity by a silt-like opening, which is guarded by the ilio-cœcal valve: this, in health, allows the transit of alimentary and excremental matter from the ileum to the colon, but not in the reverse direction. The *appendix vermiformis* is a small diverticulum which proceeds from the posterior part of the cœcum; it is the size of a goose-quill in diameter, and from three to five inches in length; its orifice of entrance into the cœcum is guarded by a small valve: sometimes foreign bodies, as bird-shot, cherry stones, &c., get into the cavity of the appendix and excite ulceration, perforation, and fatal peritonitis. The appendix is sometimes 8 or 9 inches long, and may become constricted in a hernial sac: the cœcum has no valvulæ conniventes, but is thrown into irregular sacculi by the three longitudinal bands.

The colon extends from the cœcum to the rectum, and is divided into four portions, viz.: the right or *ascending* colon, the middle, or *transverse*, the left or *descending*, and the *sigmoid flexure*.

The rectum extends from the sigmoid flexure of the colon to the anus; its upper third is wholly covered by peritoneum, its middle third is only covered by this membrane upon its anterior aspect and sides, and its inferior third has no peritoneal covering. In the male subject the antero-inferior aspect of the rectum is connected to the inferior surface of the bladder, the vesiculæ seminales, and the prostate gland, and in the female to the uterus and vagina. The rectum has the longitudinal fibres scattered over its whole surface, and is not sacculated like the other parts of the large intestine.

SALIVARY GLANDS

are three in number, viz., the Parotid, the Submaxillary, and the Sublingual.

Parotid gland, the largest of the three, is bounded superiorly by the zygoma, posteriorly by the mastoid process and sterno-mastoid muscle, and advances on the side of the face partly resting upon the masseter muscle. It sends off processes which fill the posterior part of the glenoid cavity, the fossa between the ear, ramus and angle of the lower jaw, and the intervals between the pterygoid, digastric, and styloid muscles; it often unites with the submaxillary gland. Its duct (*duct of Steno*) passes across the masseter muscle, pierces the buccinator muscle, and opens into the mouth opposite the second superior molar tooth. The portio dura nerve, the internal maxillary artery, and the external jugular vein, pass through the substance of the parotid gland. This gland cannot therefore be extirpated, without destroying the above named vessels and nerve. Two or three lymphatic glands are sometimes found on the surface of the parotid which may become enlarged and form tumours, which may be mistaken for those of the parotid itself. A small gland (*socia parotidis*) occasionally is found between Steno's duct and the zygoma, the duct of which unites with that of the parotid gland.

Submaxillary gland, placed in the digastric space, and covered by the skin, platysma-myoides muscle, and superficial fascia, is of oval figure, and much smaller than the parotid. Its duct (*Whartonian*) turns round the posterior edge of the mylo-hyoid muscle, and runs forward and inwards, upon the hyoglossus muscle, towards the *frœnum linguœ*, at the side of which it opens into the mouth.

Sublingual gland, placed between the mucous membrane of the mouth and the mylo-hyoid muscle, is the smallest of the salivary glands, and opens by several small ducts (*Rivinian*), which perforate the mucous membrane, reflected from the side of the tongue.

LIVER.

Situated in the right hypochondriac, the epigastric, and partly in the left hypochondriac regions, is the largest gland in the body. It presents an upper convex surface, a lower irregularly concave surface, a posterior thick margin attached to the diaphragm, and an anterior inferior margin which is free. The upper surface is unequally divided by the falciform ligament into a right and left lobe. The inferior surface presents the following fissures and depressions.

1. *Longitudinal fissure* extends from a notch in the anterior thin edge of the liver backwards and upwards, defining the boundary between the right and left lobes of the organ; it crosses the transverse fissure at right angles, and is continued to the superior edge of the posterior surface of the liver, by a canal often concealed in the substance of the liver. As far as the transverse fissure it contains the remains of the obliterated umbilical vein: beyond that point, the obliterated ductus venosus.

2. *Transverse fissure* extends from the longitudinal fissure into the right lobe of the liver; it contains the trunk of the right and left hepatic arteries, the trunks and branches of the portal vein, the hepatic extremities of the biliary ducts, the hepatic plexus of nerves and absorbents.

3. *Fissure of the vena cava*, situated to the right of the horizontal fissure and behind the transverse fissure, forms the right boundary of the lobulus Spigelii.

4. Depression for the gall-bladder, situated to the right of the lobulus quadratus.

5. A broad notch in the posterior thick edge of the liver, which corresponds to the right crus of the diaphragm. Besides these there are superficial depressions for the colon, right kidney, and stomach.

Lobes of the liver. 1st, right lobe, the largest; 2d, left lobe, separated from the right by the horizontal fissure; 3d, lobulus Spigelii, bounded before by the transverse fissure, and placed between the ductus venosus and vena cava; 4th, lobulus caudatus, extending from the lobulus Spigelii along the right lobe, and lying posterior to the transverse fissure; 5th, the lobulus quadratus, which is bounded behind by the transverse fissure, to the left by the horizontal fissure, to the right by the gall-bladder, its anterior edge being free.

Vessels of the liver.—1st, the hepatic artery; 2d, the vena portæ; 3d, the venæ cavæ hepaticæ; 4th, the biliary or hepatic ducts; and 5th, the absorbents. The hepatic artery is the nutrient vessel of the liver; its terminal branches open into the subdivisions of the vena portæ. The vena portæ conveys, like an artery, the blood for secretion of the liver. Its first divisions, interlobular, pass between the lobules in company with branches of the hepatic artery and biliary ducts, and form the vaginal plexuses, which apply themselves to the walls of the canals in which the vessels run: the interlobular veins open by the lobular veins into a vessel which commences in the centre of each lobule, the intralobular vein, the commencement of the hepatic veins. The intralobular vein opens into the sublobular vein, and the sublobular veins form the venæ cavæ hepaticæ. The biliary ducts commence amongst the lobular veins.

Ligaments of the liver.—Besides the liver being invested with a proper fibrous capsule, it is also covered by peritoneum, which forms four of its ligaments, viz. 1st and 2d, a right and left lateral, triangular in form, and connecting the right and left lobes to the diaphragm; 3d, a suspensory or falciform ligament, which connects its upper convex surface to the right rectus muscle, and to the diaphragm; and 4th, the coronary ligament, which connects the superior thick border of the diaphragm. There is also a fifth ligament (*ligamentum teres*), the obliterated umbilical vein, which extends obliquely from the umbilicus upwards and backwards to the anterior portion of the horizontal fissure.

Gall-bladder, pyriform in shape, and composed of an internal mucous coat, a proper fibrous coat, and a partial serous covering, is lodged in a depression on the under surface of the right lobe of the liver. Its large extremity, or *fundus*, is directed downwards and forwards; its upper extremity terminates in the *cystic duct*, about an inch and a half in length, which unites with the *hepatic duct*, formed by the union of the right and left ducts from the corresponding lobes of the liver. The common biliary duct thus formed by the cystic and hepatic ducts is called the *ductus communis choledochus*, which is about three inches in length, and conveys the bile to the duodenum.

PANCREAS.

A flattened oblong gland from five to six inches in length, similar in its formation to the salivary glands. It consists of a left *caudal extremity*, situated in the lower part of the left hypochondrium, a *body* which crosses, anterior to the left crus of the diaphragm, the aorta and the vena portæ, to the right side; and a right extremity (*the head*) which is the largest part of the gland, and which is surrounded by the duodenum. Its duct, of a whitish colour, extends along the centre of the gland from left to right, but lying near its posterior aspect; receiving the small ducts from the granules of the pancreas, it finally opens into the duodenum, close to the ductus communis choledochus, which it sometimes joins.

THE SPLEEN,

Connected to the large extremity of the stomach by blood-vessels and peritoneum, and situated between the stomach and ribs of the left side, is of a deep blue venous color, and varies in weight from six to fifteen ounces, its figure being oval. It presents a convex surface, which corresponds to the ribs, and a concave surface towards the stomach; it is enveloped by a proper fibrous membrane, and a peritoneal coat. Its interior is composed of *cells* separated by *septa*, with *white granules* intermixed. It has no excretory duct. Its function is not understood; it is supposed by some to be concerned in the formation of the red corpuscles of the blood.

URINARY APPARATUS.

THE KIDNEYS.

Two glandular bodies of oval form, situated behind the peritoneum, in each lumbar region, lying upon the diaphragm, psoas magnus, and quadratus lumborum muscles, and enveloped in a thick layer of adipose tissue. The right kidney, which is rather lower than the left, is below the liver, above the cœcum, and behind part of the duodenum and the ascending colon; the left being bounded above by the spleen, below by the sigmoid flexure of the colon, and anteriorly by the descending colon. The extremities and outer border of each kidney are convex, whilst the inner margin presents a concave aspect called *the renal fissure*, which contains the trunks of the blood-vessels and its excretory duct, observing in general the following order,—

the veins anterior, the arteries behind these, and the ureter obliquely behind both. Besides the adipose capsule which envelopes each kidney, it also has a proper fibrous coat, which adheres closely to its outer surface, and sends prolongations into its interior, as far as the calices.

The *structure* of the kidney is essentially tubular. On splitting open a kidney lengthwise we find that it evidently consists of two distinct portions, the cortical and the medullary. The cortical substance forms the whole circumference of the kidney; it is about two to three lines in thickness, of a reddish brown colour, very vascular, and from its inner surface processes pass down, separating the pyramids of the medullary portion from each other. The cortical portion consists of myriads of intricately convoluted uriniferous tubes, and of minute arteries, veins, and capillaries. An immense number of small spherical bodies are found in the cortical substances, to which the name of Corpora Malpighiana has been applied. They consist of a ball of convoluted capillaries. If we trace up a convoluted tube in the cortical substance, we find that it gradually expands, and forms a capsule, which embraces and surrounds the Malpighian body. The *medullary* portion is arranged in masses of a pyramidal shape, called the pyramids of Malpighi. Their bases are turned towards the cortical substance, their pointed extremities project free into the calices of the kidney, and are called papillæ. The number of the pyramids is from 9 to 20. They are composed of straight uriniferous tubes, converging from the base of the pyramid to its apex. They are also very vascular. The tubuli uriniferi both in the cortical and medullary portions are lined by epithelium. Those of the cortical portion lie in the midst of a plexus of capillaries, and of minute *veins*. It is believed by Bowman that the uriniferous tubes, by means of their epithelial lining, separate the urine from the blood of the vessels which are in contact with them—the water of the urine being derived from the Malpighian body, the saline portion from the plexus of minute veins.

The Calyces are small membranous sacs which surround one or more papillæ.

The Infundibula are three funnel-shaped tubes formed by the union of the calyces.

The Pelvis is the membranous reservoir formed by the union of the three infundibula.

The watery parts of the urine are secreted by the Malpighian bodies, which lie in small sacs, the commencement of the uriniferous tubules; the saline parts are separated from the blood of the venous capillaries, which convey the blood back to the renal vein.

The ureters extend from the termination of the pelvis of either kidney to the bladder. Each ureter about eighteen inches long and of the diameter of a moderate sized quill, passes behind the peritoneum, lying anterior to the psoas magnus muscles and to the iliac vessels, and gaining the inferior and posterior part of the bladder, passes obliquely between its coats and perforates its interior at the outer angle of its trigone.*

THE URINARY BLADDER.

This musculo-membranous viscus, when moderately distended, is of ovoid figure and occupies the lower region of the pelvis, behind the symphysis pubis, and anterior and superior to the rectum in the male, and the uterus and vagina in the female.

Ligaments of the bladder are divided into *true* and *false*. The true ligaments are four in number, viz., two anterior and two lateral. The anterior, white and cordlike, extend from the posterior surface of the pubes to the front of the prostate and neck of the bladder; the lateral, thinner and wider, pass from the sides of the prostate and bladder to the sides of the pelvis, and thus form the *pelvic partition*, between the parts in the pelvis, and those in the perinæum. The lateral ligaments constitute the *vesical fascia*. Both are reflections of pelvic fascia. The false ligaments are five in number, viz., two posterior, two lateral, and one superior, and are formed by the reflections of the peritoneum, some of which partially enclose the obliterated umbilical arteries, and the urachus.

---

* To gain a view of the parts just described, a perpendicular section of the gland should be made from its convex to its concave margin.

The regions of the bladder are divided into six. 1st. *The superior region*, to which are attached the urachus and obliterated umbilical arteries. 2d and 3d. *The lateral regions*, on which the vesical fascia of either side passes. 4th. *The anterior region*, the aspect of which looks towards the recti muscles, and the posterior surface of the pubes. When the bladder is distended, the anterior region rises above the pubes, and presents a surface uncovered by the peritoneum; through this region an opening may then be made to relieve retention of urine, or for the high operation of Lithotomy. 5th. *The posterior region*, the aspect of which looks towards the rectum in the male, and the uterus in the female. 6th. *The inferior region* or *fundus*, which rests on the vesiculæ seminales, the prostate gland, and the rectum in the male, and on the uterus and vagina in the female.

Coats of the bladder.—Besides the partial peritoneal covering which invests all the posterior region and the posterior parts of the upper and lateral regions, there are also three *proper* coats: 1st. *The muscular*, placed externally, the fleshy fibres of which take two directions; the external run longitudinally (the anterior and superior fibres, being stronger, have been distinguished by the name of *detrusor urinæ*), deep fibres immediately in connection with the mucous coat, are circular and best developed around the neck of the bladder. 2d. *The cellular coat*, and 3d. *The mucous*, which is exposed on opening the bladder. This coat is in general thrown into rugæ by the projection of the muscular fibres.

The trigone or *vesical triangle* is the name given to a smooth space enclosed between the openings of the ureters into the bladder, and the vesical orifice of the urethra. It is the most sensitive portion of the bladder.

The uvula is a small duplicature of the mucous membrane on the under surface of the vesical orifice of the urethra, and corresponds to the third lobe of the prostate gland.

The urethra, which terminates the urinary apparatus in the male, being more connected with the reproductive organs, we shall defer its consideration for the present.

THE PERITONEUM,

The largest serous membrane in the body, lines the parietes of the abdomen. and invests almost all the abdominal viscera; like all serous membranes, it is distinguished into two layers, a *parietal* and a *visceral*. The abdominal parietes being divided by a transverse incision corresponding to the umbilicus, the uninterrupted continuity of the peritoneum, and the different productions it forms in its course, may be thus demonstrated:—lining the inner surface of the upper section of the abdominal parietes, it ascends to the margin of the thorax, and lines the inferior suface of the diaphragm; from this muscle it is reflected on the spleen on the left side, and on the liver on the right side, forming its ligaments (vide *ligaments of liver*). From the transverse fissure of the liver, the two layers which cover the convex and concave aspects of this gland meet, and are conducted by the hepatic vessels to the lesser curvature of the stomach, thus forming the *gastro-hepatic omentum*, which contains the hepatic artery, portal vein, and biliary ducts, surrounded by a fibrous structure, called Glisson's Capsule. The artery lies to the left, the ductus communis choledochus to the right, and the portal vein between and behind. At the lesser curve of the stomach the two laminæ of the gastro-hepatic omentum separate and enclose this organ, passing from its left extremity to the spleen, thus forming the gastro-splenic omentum; at the great curve of the stomach, and lower extremity of the spleen, the two laminæ again meet, and descend in front of the colon and the small intestines to the lower part of the abdomen; they then turn upon themselves backwards, and ascend, forming the great omentum, to the transverse arch of the colon, where they separate to enclose this intestine. Having enclosed the colon, the layers again unite and form the transverse meso-colon, which passes backwards to the spine; having arrived at the spine, the two laminæ separate into a descending and an ascending layer; the descending layer passes into the lumbar regions, where it is reflected upon the ascending and descending colon, forming the right and left lumbar meso-colon; it attaches itself to the left sides of the bodies of the lumbar vertebræ, forming the anterior lamina of the root of the mesentery; from this fixed point it is continued around the jejunum and ileum, forming the peritoneal coat of these intestines, and returns again to the spine, forming the posterior lamina of the root of the mesentery. This layer of the transverse meso-colon having thus formed the mesentery, still pursues its descending course, and passes into either iliac region, and into the pelvis; on the right it attaches the cœcum to the right iliac fossa, thus forming the meso-cœcum; on the left side it attaches the sigmoid flexure of the colon to the left iliac fossa, forming the sigmoid meso-colon; and in the middle it connects the upper portion of the rectum to the upper and anterior part of the sacrum, forming the meso-rectum. Still pursuing its course downwards, and covering the upper and anterior aspect of the middle third of the rectum, it is at length reflected on the posterior surface and sides of the bladder to the superior region of this viscus, from which, and from the iliac fossa, it is reflected on the inner surface of the lower section of the abdominal parietes to the transverse incision, from the upper edge of which the description was commenced. Having thus traced the descending layer of the transverse meso-colon, the continuity of the ascending layer remains to be noticed:—ascending in front of the inferior and middle portions of the duodenum and of the pancreas, it is conducted to the diaphragm by the vena cava where it becomes continuous with the peritoneum, which has been reflected from the posterior aspect of the liver.*

Foramen of Winslow.—By this opening the cavity which is between the layers of the great omentum communicates with the general peritoneal cavity of the abdomen. It is of oval form, being bounded anteriorly by the gastro-hepatic omentum, posteriorly by the ascending layer of the meso-colon, superiorly by the liver, and inferiorly by the duodenum.

Inguinal Pouches.—As the peritoneum is ascending on the lower part of the abdominal parietes, it is thrown into four pouches, two on either side, by the obliterated hypogastric artery. The *external* pouch, between the ilium and hypogastric artery, is the largest and corresponds to the internal abdominal and the femoral rings; the *internal* corresponding to the external ring.

In the female the peritoneum passes from the rectum on the upper and back part of the vagina, from which it ascends on the uterus, forming on each side its broad ligaments, and is reflected from the anterior part of the uterus to the back of the bladder.

---

* This, the usual description of the peritoneum, leaves unexplained the way in which the hepatic vessels reach the liver without perforating the membrane. This point can be understood only by referring to the history of the development of the fœtus, in which the intestinal tube, nearly vertical, is bound to the spinal column by two folds of peritoneum, between which lie the aorta and its branches. When the viscera assume the position known in the adult, it is impossible to trace all the peritoneal folds.

MALE ORGANS OF GENERATION.

THE TESTICLES

are two in number, of oval form, are contained in the scrotum, and are likewise enveloped by proper tunics.

Tubuli seminiferi are numerous small tubes which form the body of each testicle. They are very long and tortuous, and are arranged in conical lobules, which are separated from each other by fibrous bands, derived from the tunica albuginea. Lauth states the number of tubes to be 840, and the length of each to be 27 inches.

Vasa recta, from sixteen to twenty in number, are formed by the union of the tubuli seminiferi, and are contained between the layers of the corpus Highmorianum. The *vasa recta* terminate in a kind of plexus or network called the *rete testis*. From the *rete testis* proceed from 12 to 18 ducts, called *vasa efferentia*. These are so convoluted as to form the *coni vasculosi*. The *coni vasculosi* form the *globus major*, or head of the *epididymis*, which is situated at the upper part of the body of the testicle; still convoluted, it is traced downwards, forming the body of the epididymis, which is narrow, and placed at the posterior part of the body of the testicle, and arriving at the inferior part of the glands it forms the *globus minor* (or tail of the epididymis). The vas deferens having thus formed the epididymis, escapes from the globus minor, and having increased in size and density, ascends along the inner aspect of this body, until it becomes connected to the spermatic vessels and cremaster muscle; it then passes through the external abdominal ring and the inguinal canal, and having passed through the internal abdominal ring, it separates from the spermatic vessels, and is conducted by the false lateral ligaments of the bladder to this viscus, along the side and inferior part of which it runs, lying internal to its corresponding vesicula seminalis. It here approaches its fellow of the opposite side, and both ducts becoming flattened arrive at the base of the prostate gland, where they are joined by the ducts of the vesiculæ seminales, their union forming the *common ejaculatory ducts;* these run through the prostate gland, and open into the prostatic portion of the urethra, at the side of the verumontanum. The canal of the vas deferens is small. Surrounding this canal, are organic muscular fibres, by the action of which the semen can pass upwards from the testicle against the force of gravity.

PROPER COATS OF EACH TESTICLE.

Tunica albuginea.—A strong fibrous investment, of opaque white colour, which forms the capsule of the gland. From its inner surface it sends two laminæ, which project into the back part of the testicle for about two lines, forming the body called *corpus Highmorianum:* from the free edge of this proceed the fibrous bands, already mentioned as separating the conical bundles of tubuli seminiferi.

Tunica vaginalis.—A serous membrane, consisting of two layers, one covering the testicle, called *tunica vaginalis testis*, the other lining the scrotum, called *tunica vaginalis scroti.* When the tunica vaginalis scroti is divided, its continuity with the visceral layer may be demonstrated by tracing the membrane, when it will be found to be reflected on the side and fore part of the epididymis and testicle, forming a pouch between these bodies, and also for a short distance on the fore part of the chord. Hydrocele is a collection of serum in the tunica vaginalis.

Tunica communis, formed by the fibres of the cremaster muscle and cellular membrane, surrounds the chord and the fore part and sides of the testicle.

COMMON COVERINGS OF BOTH TESTICLES.

The scrotum, a prolongation of the common integument, is of brownish color, slightly studded with hairs and sebaceous follicles, presenting in the median line a hard ridge, called the *raphe*, from either side of which it is thrown into rugæ.

The dartos, composed of elastic tissue, mixed with unstriped muscular fibres, is formed by the subcutaneous cellular tissue and the ramifications of numerous bloodvessels, which were formerly supposed to give this coat a reddish appearance. The vermicular or rolling motion of the testicles is caused by the action of the dartos.

The Superficial fascia lies immediately under the dartos, is derived from the superficial fascia of the abdomen, and is continuous with the superficial fascia of the perineum.

Septum scroti.—This partition, which divides the scrotum into two, is formed by the dartos and superficial fascia, these membranes being attached to the raphe, and from thence ascending between the testicles to the urethra.

Each testicle is supplied with blood by the spermatic artery, the blood of which is returned by the spermatic veins; it receives nerves from the spermatic plexus, which is formed by branches from the renal and aortic plexuses of the sympathetic.

The spermatic cord is composed of the vas deferens, with its small artery, derived from the vesical; the spermatic artery, veins, and nerves; the genito-crural nerve, the ilio-scrotal nerve, an artery to the cremaster coming off from the epigastric and absorbents; all of which are connected to each other by fine cellular tissue, and are enveloped by fascia and the cremaster muscle. The chord, thus formed, extends from the epididymis to the internal abdominal ring, where its constituents separate from each other.

Corpus pampiniforme is the name given to the venous plexus formed by the spermatic veins after these vessels have escaped from the testicles. The supermatic vein on the right side enters the vena cava at an acute angle; on the left side, it joins the renal vein at a right angle. This may be given as one reason why varicocele is more common on the left side.

The vesiculæ seminales are two oblong flattened bodies, situated at the inferior fundus of the bladder, behind the prostate gland, and on the outer side of the vasa deferentia. Each seminal vesicle is about two inches long, and consists of a long tortuous membranous tube convoluted on itself, the small excretory duct of which joins its corresponding vas deferens.

The prostate gland is a flat conoidal body, the base being posterior, corresponding to the vesiculæ seminales, the apex anterior, corresponding to the vesical extremity of the urethra. It is divided into three lobes; two lateral, large, and united in the mesial line, their union being marked by a slight groove; and a third or small lobe, situated in the angle between the two lateral lobes towards the base of the gland. The prostate gland is firm and resisting to the touch, and composed of numerous follicles, with minute ducts, which unite to form larger tubes, the openings of which, ten or twelve in number, are on the under surface of the urethra, on either side of the verumontanum. Posterior to the base of the prostate is a triangular space, bounded on each side by the vasa deferentia, anteriorly by the prostate, and posteriorly by the reflection of the peritoneum. In this space the bladder may be punctured through the rectum in certain cases of retention of urine.

Cowper's glands are two small oblong-round bodies, placed before the prostate gland, and immediately behind the bulb. They lie below the layers of the triangular ligament. The duct of each gland having run a course of about an inch, opens into the urethra a little anterior to its bulb.

THE PENIS.

This organ consists of two long cylindrical bodies, named corpora cavernosa, and a body named corpus spongiosum, which contains the urethra, all these parts being connected together and surrounded by the superficial fascia and the common integuments.

The corpora cavernosa are two cylindrical bodies, united to each other in the mesial line. They are composed of erectile tissue, vessels and nerves, surrounded by a dense fibrous covering.

Each *corpus cavernosum* commences by the *crus penis*, which is the narrowest part, and which is attached to the rami of the ischium and pubes, covered by the erector penis muscle. At the symphysis pubis both crura unite, forming the chief part of the body of the penis, and terminate anteriorly in an obtuse point, to which is intimately attached the glans penis.

Septum pectiniforme, a partition, imperfect, as its name implies, which corresponds to the mesial line, and marks the division of the corpora cavernosa.

The urethra is a membranous canal, extending from the neck of the bladder to the extremity of the glans penis, its length and width varying according to the erect or collapsed state of the organ. In the latter condition its length is from seven to eight inches long, and its calibre about three or four lines. It is lined by mucous membrane, and is distinguished into, 1st, the prostatic portion, which is an inch and a quarter in length; 2d, the membranous portion, which is about three quarters of an inch long; and 3d, the spongy portion, which occupies the remainder of its length.

The corpus spongiosum urethræ is a cellulo-vascular tube surrounding the urethra, and occupying the under mesial line of union of the corpora cavernosa; it commences in the bulb of the urethra, and extends along the canal to its extremity, where it terminates in the glans penis, the bulb and glans penis being merely expansions of this structure.

Upon exposing the mucous surface of the urethra by an incision, we observe, 1st, a slit-like contraction at the orifice; 2d, behind this a dilatation, called *fossa navicularis*; 3d, the constant diameter of the canal until we arrive at the bulb, where it becomes gradually and very slightly dilated; forming, 4th, the *sinus of the bulb;* 5th, the narrowest part of the canal, which corresponds to the membranous portion; 6th, the dilatation corresponding to prostate gland; and 7th, a contracted orifice at its termination in the bladder. In the prostatic portion of the urethra, a prominent fold of mucous membrane, called *verumontanum*, projects from its under surface, and presents in its centre a large lacuna, the *sinus pocularis*, the orifice of which is directed forwards. The verumontanum is the point to which Lallemand's caustic is to be applied, in cases of spermatorrhœa, &c., attended with great irritability of the urethra. On either side of the verumontanum the prostatic sinuses are situated. Upon the upper surface of the urethra, from the orifice to the bulb, are the openings of numerous mucous follicles, directed forwards, the largest of which is about an inch from the orifice, and is called, from its size, *lacuna magna*. This, with the other mucous follicles, is often the seat of obstinate gonorrhœa. The ducts of the mucous follicles open on the surface of the urethra: the orifices of Cowper's glands open a little anterior to the sinus of the bulb, the common ejaculatory ducts on either side of the verumontanum, and the ducts of the prostate in the prostatic sinuses.

The superficial fascia, which envelopes the penis, is derived from that of the abdomen, and terminates at the corona glandis. It is strong where it passes from the linea alba upon the dorsum of the penis, forming the *suspensory ligament*, but is exceedingly delicate and loose upon the body of the organ.

The skin of the penis is remarkably thin and loose, and extending for an indefinite length beyond the organ, is reflected inwards, and intimately attached to the corona glandis; the loose fold thus formed being called the *prepuce*. From the corona glandis it is continued along the glans until it becomes identified with the mucous membrane at the orifice of the urethra, having first formed the fold which lies posterior and inferior to this opening, called *frœnum preputii*.

Glandulœ odoriferœ are a number of small sebaceous glands which surround the corona glandis, and which lie beneath the skin.

THE FEMALE ORGANS OF GENERATION.

The ovaries are two ovoidal bodies, placed, one on either side of the womb, in the duplicatures of the peritoneum, called the broad ligaments of the uterus. Each ovary, enveloped by a white fibrous membrane, consists of a pulpy brownish grey substance, highly vascular, and containing from fifteen to twenty minute vesicles, each of which is composed of a thin membrane containing a viscid yellowish fluid; these are called the *Graafian vesicles*.

The Fallopian tubes are the excretory ducts of the ovaries; each is about four inches in length, and is contained in the broad ligament, one extremity being attached to the superior angle of the uterus into which it opens by a small orifice (*orificium uterinum*), the other being free, and surrounded by a fringe (*corpus fimbriatum*), in the centre of which is the peritoneal aperture (*orificium superius*). At this point the mucous membrane of the Fallopian tube communicates with the serous membrane, the peritonæum, forming the only exception to the rule, that the serous membranes are perfect bags or sacs, and have no opening.

The uterus is a hollow organ of pyriform shape, and is distinguished into the *fundus*, the *body*, and *cervix*. The fundus is superior and posterior, and receives at either angle the Fallopian tube : the body is intermediate between the fundus and the neck, the latter being inferior and anterior, and surrounded by the vagina : at the extremity of the neck is a small elliptical opening, surrounded by a thick margin, which, from its resemblance to the mouth of a tench, has been called *os tincæ*, as well as *os uteri*. The cavity of the uterus is small compared to the thickness of its walls, and is of triangular shape; its superior and outer angles presenting the orifices of the Fallopian tubes, the inferior angle presenting the os tincæ. The uterus is placed between the bladder and rectum.

The vagina is a membrano-vascular tube, extending from the neck of the uterus to the external outlet, where it is continuous with the surface. It is composed of mucous membrane, surrounded by cellular tissue, a vascular network, and the sphincter vaginæ muscle; its length is about four or five inches, its breadth one; but being very distensible these measurements vary. Its lining membrane is thrown into transverse rugæ on its anterior and posterior surface, and is studded with the orifices of numerous mucous follicles. The colour of the membrane varies, at the external orifice being red, and of a grey and sometimes marbled colour as it approaches the uterus.

The mons veneris is a soft adipose eminence, situate on the upper and anterior part of the pubes, covered by common integument, which in the adult is thickly set with hairs.

The vulva is the fissure extending from the mons veneris to the perineum.

The labia magna are large folds of the integuments which bound the vulva on either side, and unite below in a crescentic edge (*the fourchette*). They contain large mucous glands about the centre.

The clitoris, a small oblong conical body, placed between the upper extremities of the labia. It consists of a structure similar to the corpus cavernosum in the male, and arises by two crura from the pubes; these unite to form its body, at the extremity of which is placed a red protuberance, called the *glans clitoridis*, over which is thrown a loose fold of integuments (*the prepuce*).

Meatus urinarius is about half an inch below the clitoris.

Labia parva, or nymphæ, are two red crescentic folds of mucous membrane, enclosing erectile tissue; they descend, one on each side, from the prepuce of the clitoris, and are lost about the centre of the vulva.

The Hymen, when it exists, is a crescentic fold of mucous membrane, surrounding the sides and inferior orifice of the vagina.

The carunculæ myrtiformes are small reddish bodies surrounding the orifice of the vagina; they are sometimes described as the remains of the hymen.

THE MAMMÆ

are two in number, situated at the anterior and superior part of the thorax, and connected to the great pectoral muscles by a capsule of condensed cellular tissue. Each of these glands, of a hemispherical shape, consists of vessels and numerous lactiferous tubes, arising from dilated blind extremities, or cells: the tubes are grouped together to form lobes and lobules; as they approach the nipple they become considerably dilated, and form sinuses, but in the nipple they are again reduced in size, and terminate at the apex by open orifices surrounded by delicate muscular tissue. The *nipple* is a conical process, surrounded by a brownish areola, and composed externally of the integuments, which are very thin, and internally of the lactiferous tubes, together with numerous blood-vessels, from which the nipple derives its property of occasional erection.

ORGANS OF THE SENSES.

THE ORGAN OF TOUCH.

The skin is composed of the cuticle, or epidermis, the rete mucosum, and the corion or cutis vera.

The cuticle is composed of epithelium cells, of which the more superficial are flattened and dried; they are deposited in thickest layers upon the soles and palms.

The deepest layer of cells of the cuticle, or that which lies immediately upon the cutis vera, is colored, in the negro and other dark-skinned races. To this colored layer the name of *Rete Mucosum* has been applied. The cuticle with its pigment is removed by maceration.

The corion is a dense strong membrane, consisting of fibres interwoven with each other, which are more firmly compacted the nearer they are to its outer surface. Its internal surface is cellular, its external very vascular, and presenting numerous small conical papillæ; at the extremities of the fingers these papillæ are best developed, are furnished with minute nervous filaments and covered with very thin cuticle; thus affording a delicacy of organization necessary for the greater perfection of the sense of touch.

The arrangement of the blood-vessels varies according to the delicacy of the sense of touch.

The capillaries of the papillæ end in loops.

The skin is studded with minute hairs, and with sudoriparous and sebaceous glands.

The sudoriparous or sweat glands consist of a convoluted tube, which opening upon the surface of the cuticle, may be traced downward through the skin, until it end in a *coil*, forming a small rounded body in one of the meshes of the corium.

THE ORGAN OF SMELL.

The nose is bounded superiorly by the nasal, frontal, ethmoid, and sphenoid bones; inferiorly by the palatine plates of the superior maxillary and palate bones; externally on either side by the superior maxillary, lachrymal, inferior spongy, ethmoid, and palate bones, and by the internal pterygoid plates of the sphenoid bone. It is divided into the *two nares* by the *septum nasi*, which is formed by the azygos process of the sphenoid bone, the nasal plate of the ethmoid bone, the vomer, and the mesial spines of the superior maxillary and palate bones. Besides the bony boundaries, the nose presents, anteriorly, five cartilages, which form the *anterior nares*, or the nostrils. The middle verticle cartilage is of triangular form, and rests in the fissure of the vomer inferiorly, is attached to the vertical plate of the ethmoid bone above, and presents anteriorly a subcutaneous, free, thick edge, and thus completes the septum nasi. The lateral cartilages which form the wings of the nose are also triangular, are attached to the superior maxillary and nasal bones, and in the median line to the vertical cartilage. The inferior lateral fibro-cartilages are attached to the three cartilages just described, are thick and semicircular, forming, with the vertical cartilage, the anterior-inferior oval openings of the nostrils.

The posterior nares are of oval shape, and open into the upper part of the pharynx; they are separated from each other by the posterior free edge of the vomer, are bounded superiorly by the body of the sphenoid bone, inferiorly by the palate bones, and externally by the internal pterygoid plates of the sphenoid bone. The external lateral wall of each nostril, from the arrangement of the spongy bones, forms three fossæ, called *meatuses*, with which several orifices communicate.

In the inferior meatus, at the junction of its anterior with its middle third, is the opening of the *nasal duct*, and posteriorly, on a level with the inferior spongy bone, is the opening of the Eustachian tube. In the middle meatus is the slit-like opening of the antrum maxillare, anterior to which is the groove called *infundibulum*, which leads from the frontal sinus, and into which open the anterior ethmoidal cells.

Into the superior meatus, the posterior ethmoidal cells and the sphenoidal sinus open. The interior of the nose is lined with the Schneiderian membrane, which is highly vascular and sensitive, and consists of two layers; a fibrous layer, which is the periosteum, or perichondrium

cavities, and a mucous membrane.
which supply the nasal cavities are
(which pass through the cribriform
ethmoid bone), the internal nasal of
nic, and branches derived from
iglion.

THE ORGAN OF TASTE.

ie presents several papillæ covered
iembrane. It is of triangular form,
l by its base to the os hyoides, by
ous membrane to the epiglottis and
by muscles to the lower jaw. It is
lar, and receives six nerves, three on
iz., the gustatory branch of the fifth
e ninth, or lingual, for motion, and
aryngeal, the function of which is
t most probably is connected with
taste.
æ are of three orders,—the filiform,
le tip and sides; the fungiform, scat-
he dorsum; the circumvallatæ at

THE ORGANS OF VISION

ie globes of the eyes and their ap-
The eye-ball is composed of mem-
luids, called humours.
ic coat, occupying about four-fifths
, is a strong fibrous structure, and
id and anteriorly than in its centre.

Its outer surface is in contact with the adipose tissue of the orbit, the tendinous expansions of the muscles of the eye, and anteriorly with the conjunctiva; its inner surface is lined by the choroid coat. It presents posteriorly a small aperture for the transmission of the optic nerves, and an anterior large one, about six or seven lines in diameter for the cornea.

The cornea, which forms the anterior fifth of the globe, is smooth and transparent. It consists of three layers, viz. the conjunctival layer externally, an elastic layer internally, and between both, the proper cornea, which is composed of fibres connected together by fine cellular tissue.

The *choroid coat* is a thin vascular membrane, situated between the sclerotic coat and the retina; it extends from the entrance of the optic nerve to the ciliary ligament, to which it is firmly connected; it is then directed inwards, and forms the folds called ciliary processes. Its internal surface is coated by a dark brownish secretion, called *nigrum pigmentum*, its outer surface being connected to the sclerotic coat by fine cellular tissue, and by the ciliary vessels and nerves; on this surface the veins observe an arched arrangement, and are called *vasa vorticosa*.

The *ciliary ligament* is about a line and a half in breadth, of a greyish white cellular structure, and corresponds to the circle of junction of the cornea and sclerotic coat externally, and of the choroid and iris internally.

The *ciliary processes* vary in number from sixty to seventy, and are productions or continuations of the choroid coat; each ciliary process is of a triangular figure, the anterior edge being attached to the ciliary ligament, the posterior to the hyaloid membrane, and the internal free projecting into the posterior chamber of the aqueous humour, towards the lens, but not attached to this body.

The *Iris* is a circular membrane placed in a transverse vertical position, attached by its larger circumference to the ciliary ligament, floating in the aqueous humour, and presenting a circular opening in its centre called the *pupil*. It divides the space between the anterior surface of the capsule of the lens, and the posterior surface of the cornea, unequally, into what are termed the *chambers of the aqueous humour*, the anterior chamber being much the largest; both chambers, however, communicate through the pupil. The anterior surface of the iris presents a radiated appearance, and varies in colour in different individuals; the posterior surface is covered by nigrum pigmentum, and has received the name of *uvea*. The iris is supplied by the ciliary nerves and vessels.

The *retina*, placed between the choroid coat and vitreous humour, consists of three layers, an external or *serous layer*, called from its discoverer *membrana Jacobi*,* an internal or vascular

---

* The membrana Jacobi is not a serous membrane, but consists of rod-shaped bodies possessing high power of refraction.

layer, and betwen both the nervous layer. About two lines on the temporal side of the entrance of the optic nerve the retina presents a small hole surrounded by a yellow margin called *the foramen of Soemmering,* round which the retina is thrown into a fold.

The *aqueous humour* is contained in the anterior and posterior chambers of the eye, is perfectly transparent, and is from four to five grains in quantity.

The *vitreous humour* occupies about the three posterior fourths of the eye; it is contained in the *hyaloid membrane,* which not only envelopes it, but sends numerous partitions from its inner surface to form cells in which this transparent fluid is deposited. The vitreous humour thus contained in its capsule is convex posteriorly and on its lateral circumference, but presents anteriorly a concavity for the reception of the crystalline lens; around the circumference of this cavity the vitreous humour presents a striated appearance, caused by the marks of the ciliary processes, to which the term *corona ciliaris* is applied.

The *crystalline lens,* enclosed in its capsule and placed in the anterior depression of the vitreous humour, is a transparent body, presenting an anterior and posterior convex surface, the latter being the more prominent. Its external surface is soft and pulpy, gradually increasing in density towards its centre.

The *capsule* of this body, like the lens itself,

is transparent, and composed of homogeneous membrane.

The lens and its capsule derive their nutriment from the vessels of the retina. Opacity of the lens constitutes cataract.

Liquor Morgani is a fluid found some hours after death between the lens and its capsule.

Canal of Petit.—The lens is retained in its situation by the hyaloid membrane, which, splitting into two laminæ at its circumference, pass one anterior the other posterior to its capsule; a triangular canal is thus formed, which is intersected by minute septa; this is the canal of Petit, and may be demonstrated by distending it with air, when it will present a vesicular appearance.

THE APPENDAGES OF THE EYE.

The *lachrymal gland*, placed in the upper and outer part of the orbit, behind the external angular process of the os frontis, and about the size of a small almond, is of greyish colour, consists of numerous granules united by an imperfect capsule, and pours forth its secretion of tears by means of five or six minute ducts, which open behind the upper eyelid, in the angle formed by the reflection of the conjunctiva.

Tunica conjunctiva, a mucous membrane which lines the interior of each eyelid, and is reflected on the anterior part of the globe of the eye. At the inner angle of the eye it forms a small fold called *plica semilunaris*, covers the *caruncula lachrymalis*, and having lined the lachrymal sac and duct becomes continuous with the mucous membrane of the nose. The epithelial layer of this membrane is continued over the cornea.

Caruncula lachrymalis, a small vascular body composed of mucous glands and cellular tissue situated in the nasal angle of the eye, and covered by the membrana conjunctiva.

The *palpebræ* or *eyelids*, semicircular in form, are composed of skin externally, which is very fine, the tunica conjunctiva internally, and between both the orbicularis palpebrarum muscle, the tarsal cartilages and their ligaments, and the Meibomian glands, together with blood-vessels, nerves, and absorbents.

The superior eyelid, besides being the largest, has also peculiar to it, the levator palpebræ superioris muscle.

The *tarsal cartilages* are thin cartilaginous plates of a semicircular form, the superior being the largest; to their convex margins are attached the *palpebral ligaments*.

The *Meibomian glands*, of a yellow colour, are very numerous, particularly in the upper eyelid, and are arranged in vertical rows.

The opposed edges of the eyelids are thick, and are bevelled off obliquely towards the eye, so that when closed they only touch at their anterior edges, thus leaving a triangular canal, the base of which is formed by the tunica conjunctiva, along which the tears are conducted to the puncta lachrymalia.

The *cilia* or *eye-lashes*, attached by their roots to the opposed margins of the eyelids, observe a curved arrangement, their convexities looking toward each other.

The *puncta lachrymalia* are the two minute orifices of the lachrymal canals, placed within two or three lines of the nasal terminations of the eyelids and upon their opposed edges.

The *lachrymal canals*, the superior of which is longer and curved, the inferior being nearly straight, lead from the puncta lachrymalia to the lachrymal sac.

The *lachrymal sac*, placed in the fossa formed by the lachrymal and superior maxillary bones, behind the fibrous expansion from the tendon of the orbicularis palpebrarum, is of oval form, receives the lachrymal secretion by the puncta lachrymalia, and transmits it to the nasal duct, with which it is connected inferiorly.

The *nasal duct*, enclosed in a bony canal formed by the lachrymal, superior maxillary, and inferior spongy bones, passes obliquely downwards, backwards, and outwards, and terminates in the inferior meatus of the nose.

THE ORGAN OF HEARING.

This organ consists of the external ear, including the auricle and meatus auditorius externus;

the middle ear, including the cavity of the tympanum and its appendages; and the internal ear or labyrinth, including the vestibule, semicircular canals, and cochlea.

The *external ear* consists of a fibro-cartilaginous plate covered by skin, and so moulded as to form different elevations and depressions, which have been described with more minuteness than they deserve.

The *helix* is the semicircular eminence which forms the outline of the external ear.

The *antihelix* commences superiorly by two roots, which enclose a fossa (*fossa navicularis*), and is situated inferior to the helix.

The *tragus* is an eminence placed anterior and inferior to the meatus externus.

The *antitragus* is a smaller eminence posterior to the meatus externus.

The *lobule* is a pendulous body placed underneath the antitragus.

The *concha*, a deep conoidal cavity which leads to the meatus externus, and in which the several depressions formed by the eminences just described terminate.

The *meatus externus* is a curved canal which leads from the concha to the membrana tympani; it is lined by skin, beneath which are placed small glands (*glandulæ ceruminosæ*), which secrete the ear-wax (cerumen). The inner half of this canal is surrounded by bone.

The *membrana tympani*, separating the external from the middle ear, is of oval form, and consists of three layers, viz. the external or cuticular, the internal or mucous, and between both a fibrous layer. To its inner aspect is attached the crus of the malleus, which, by drawing it towards the middle ear, gives it a concave aspect externally.

The *middle ear* consists of the cavity of the tympanum and the small bones of the ear and their muscles.

The *cavity of the tympanum* is an irregular cylindrical space, closed externally by the membrana tympani, and bounded interiorly by a bony partition which separates it from the labyrinth. It presents the following eminences and foramina, viz. the *promontory*, a convex eminence situated on its internal side, and which marks the situation of the vestibule; the *foramen ovale*, placed above the promontory, and to which the base of the stapes is affixed; the *foramen rotundum*, below the promontory, closed by a membrane (*lesser tympanum*) which separates the scala tympani of the cochlea from the cavity of the tympanum; *the opening of the mastoid cells*, situated posteriorly and superiorly; the *pyramid*, a bony projection placed below the opening of the mastoid cells, hollow within and containing the stapedius muscle; a *small foramen* below the pyramid for the transmission of the chorda tympani nerve; anteriorly the openings of the two bony canals, the superior of which lodges the tensor tympani muscle, the inferior forming the bony part of the Eustachian tube; inferiorly is the opening of the Glasserian fissure, and superiorly are several smaller foramina for blood-vessels.

The *bones of the ear* are three in number, very small, and contained within the cavity of the tympanum.

The *malleus* is divided into the head, which is smooth and articulates with the incus; the neck, which is small, and connects the head to the shaft; *the handle or shaft*, which descends from the neck, and is attached to the membrana tympani; and the *processus gracilis*, which passes from the neck to the Glasserian fissure.

The *incus* is divided into its body, which presents a cup-like cavity for the head of the malleus; a superior crus, which is short and lies in the mastoid cells; and a long crus, to the extremity of which is attached a small process of bone, considered by some as a distinct bone, and called *os orbiculare*.

The *stapes* presents a small head, which is attached to the orbicular process; a short neck; two curved crura, which terminate in the base; and the base itself, which is of oval shape and connected to the foramen ovale.

The *internal ear* or *labyrinth* contains—

1. The *vestibule*, placed behind the cochlea and before the semicircular canals. It is a small oval cavity lined by a membrane common to the labyrinth, contains a watery fluid, and presents the following openings; viz. the foramen ovale, the five orifices of the semicircular canals, the

orifice of the scala vestibuli of the cochlea, and the orifice of the aqueduct of the vestibule.

2. The *semicircular canals*, placed behind the vestibule, are three in number, two vertical and one horizontal; of the former, one is superior, and the other posterior. The openings of these canals are only five in number, in consequence of one opening of the vertical canals being common to both.

3. The *cochlea*, of conical form, the base towards the internal meatus, the apex towards the carotoid canal, is composed of a bony tube which makes two turns and a half round a central pillar called *the modiolus*. This tube is divided longitudinally by a thin plate, half bony half membranous, called *lamina spiralis*, into two independent cavities; the two tubes thus formed are called the scalæ of the cochlea, they both unite at the apex in a cavity called *infundibulum*, and at the base of the cochlea they separate, one called *scala vestibuli*, which opens into the vestibule, the other called *scala tympani*, which opens into the tympanum by the foramen rotundum. From the scala tympani proceeds a narrow bony canal called *the aqueduct of the cochlea*, which terminates in a slit-like opening in the inferior border of the petrous bone.

4. The *auditory nerve* gains the internal ear by the minute foramina at the base of the meatus auditorius internus, and is expanded in the form

of soft pulpy filaments in the cochlea and vestibule.

THE ABSORBENT SYSTEM.

Comprehends—1st, the vessels which convey the lymph and chyle into the veins, and 2d, the enlargements which occur in their course called glands or ganglia.

The *lacteal or chyliferous vessels* commence on the villi of the mucous surface of the intestines, pass through the mesentric glands backwards towards the spine, where they terminate in the thoracic duct.

The *lymphatic vessels* are found in most situations of the body, and generally observe a deep and superficial arrangement.

Lymphatics of the lower extremities.—The superficial set accompany the external and internal saphena veins: they communicate freely in their course with the deep lymphatic trunks which accompany the deep vessels. Those which accompany the external saphena vein enter the glands in the popliteal space, whilst those accompanying the internal saphena vein ascend to the groin and pass through the inguinal glands, having formed numerous connections with the superficial lymphatics of the abdomen, the perineum, and the genitals. The deep lymphatics of the hip and perineum are conducted by the branches of the internal iliac vessels into the pelvis, and pass through the pelvic glands. From the inguinal and pelvic glands the lymphatics pass through the iliac vessels to the receptaculum chyli.

The Thoracic Duct.—This canal commences by a dilatation called *receptaculum chyli*, placed on the body of the second lumbar vertebra; passing between the crura of the diaphragm it gains the posterior mediastinum, where it lies between the aorta and the vena azygos; at the fourth dorsal vertebra it crosses the spine obliquely to the left side, passing behind the œsophagus and arch of the aorta, and placed behind the left pleura and between the left carotid and left subclavian arteries: it is then conducted by the œsophagus to the left side of the neck as high as the sixth cervical vertebra, where, making a slight curve downwards and outwards, it opens close to the external angle formed by the left subclavian and jugular veins.

Lymphatics of the upper extremities.—The superficial set accompany the superficial veins, and pass through two or three glands situated at the inner condyle; having joined the deep lymphatics which accompany the venæ comites, they proceed onwards to the axilla, and pass through the axillary glands; following the course of the axillary vein, they pass beneath the clavicle, join the lymphatics of the neck, and terminate in the thoracic duct. The lymphatics of the right upper extremity and right side of the neck unite to form the *right or lesser thoracic duct,* which opens into the right vena innominata.

The lymphatics of the trunk consist of a deep and superficial set; in the chest the former are seated between the muscles and pleura, in the abdomen between the muscles and peritoneum, the superficial being subcutaneous. The viscera contained in the chest and abdomen also have a superficial and deep layer of lymphatics, the deep being distributed through the peculiar tissue of each organ, the superficial running beneath the membranous envelope.

Lymphatics have been seen in the membranes, but not in the proper substance of the brain and spinal cord.

PECULIARITIES OF THE FŒTUS.

The principal anatomical peculiarities of the fœtus, by which it is distinguished from the adult, are the following:—

The *thymus gland* occupies the anterior mediastinum,—the *kidneys* are lobulated, and each is covered by a cellulo-vascular body called *Renal capsule,* which is larger than the kidney itself,—the *liver* is very large, particularly its left lobe,—the *lungs* are compact, of a deep red colour, and sink in water, the bronchial tubes and their ramifications being void of air,—the auricles of the heart communicate with the *foramen ovale,*—at the bifurcation of the pulmonary artery an arterial trunk about nine lines in length, called *ductus arteriosus*, proceeds to the aorta, into which vessel it opens—the *umbilical vein* proceeds to the liver, where having distributed some branches to its left lobe, it divides into the *communicating branch*, which unites into the portal vein, and the ductus venosus, which opens into the vena cava inferior,—the *internal iliac arteries*, under the name of *umbilical* or *hypogastric*, turn upwards and forwards along the sides of the bladder, pass through the umbilicus, and run a tortuous course along the umbilical vein to the placenta,—and the urinary bladder is in the abdominal part of the pelvis, from the summit of which a ligamentous cord, called *urachus*, passes to the umbilicus. Until the seventh month the pupil is closed by a membrane, called *membrana pupillaris*, and in the male the *testes* are contained in the abdomen.

CERVICAL FASCIA.

The superficial fascia is thin and consists of two layers, between which are placed the fibres of the platysma myoides. The deep fascia binds down and invests the muscles of the neck. It is a strong, dense, pearly white structure, attached behind to the spines of the cervical vertebræ, in front to the mesial line, and below to the clavicle and sternum; above it is connected with the jaw and parotid gland; and it sends a process from the styloid process to the angle of the jaw known as the stylo-maxillary ligament. The sterno-mastoid, the omo-hyoid, and the subclavius muscles, receive complete sheathes from it.

The carotid artery, pneumogastric nerve, and the internal jugular vein, with its accompanying chain of lymphatic glands, are contained in a sheath, which lies *underneath* the deep cervical fascia. These glands sometimes become greatly enlarged, and form either benign or malignant tumours, which are bound down firmly by this strong, resisting fascia.

THE FASCIÆ.

SUPERFICIAL FASCIA OF THE ABDOMEN

passes downwards from the thorax over the abdominal muscles and Poupart's ligament to the thigh. In the median line it passes off the pubes upon the penis, forming its suspensory ligament, and in the female it descends into the labia. In the male it passes on either side round the spermatic cord into the scrotum, and becomes continuous with the fascia of the perinæum. After having passed over Poupart's ligament it forms envelopes for the inguinal glands and adheres to the fascia lata, presenting a cribriform appearance (vide Fascia lata); and continuing its course downwards becomes identified with the subcutaneous cellular tissue of the lower extremities.

FASCIA TRANSVERSALIS AND FASCIA ILIACA.

The *fascia transversalis* is placed between the transversalis muscle and the peritoneum; it is very strong inferiorly, and is connected to the internal lip of the ilium and to the whole length of Poupart's ligament, and is continuous, behind the rectus muscle, with the fascia of the opposite side. As the external iliac vessels are passing beneath Poupart's ligament, a production of this fascia extends along the anterior aspect of their sheath, and becomes identified with the cribriform fascia in the groin. The spermatic cord in the male, and the round ligament in the female, pass through a foramen in this fascia about half an inch above Poupart's ligament, and midway between the spine of the ilium and the symphysis pubis; this opening is the internal abdominal ring; from its margin is prolonged over the cord a funnel-shaped process, called the infundibuliform fascia. The fascia transversalis forms a covering in all the varieties of abdominal hernia.

The *fascia iliaca* is much stronger than the fascia transversalis; it is connected to the inner lip of the ilium, passes over the iliacus internus muscle, adheres to Poupart's ligament, from which it passes behind the sheath of the femoral vessels into the thigh, and is connected with the capsule of the hip-joint and the pubic portion of the fascia lata. The processes of fascia transversalis and fascia iliaca, passing one in front and the other behind the femoral vessels, and uniting at the outer and inner border, form, the sheath of the vessels. Femoral hernia is covered by the sheath of the vessels here described. The fascia iliaca continued into the pelvis beomes the pelvic fascia; it lines the parietes of this cavity as far as the upper origin of the levator ani muscle, where it divides into two layers; one layer (the outer) called the *obturator fascia*, descends between the obturator internus muscle and the levator ani, and is inserted into the great sciatic ligament, the tuberosity of the ischium, and pubes. The *internal* layer of the pelvic fascia, called also *vesical fascia*, passes downwards along the inner surface of the levator ani muscle to the inferior margin of the symphysis pubis, *from which it is reflected on the prostate gland and neck of the bladder*, forming the *anterior true ligament* of the bladder, and *laterally* it is reflected on the sides of this viscus, forming its *true lateral ligaments*. This vesical fascia passing from the side of the prostate and bladder to the side of the pelvis, forms the "pelvic partition."

SUPERFICIAL PERINÆAL FASCIA

strongly adheres to the rami of the ischium and pubes of either side, and extends across the perinæum, being continuous anteriorly with the superficial fascia of the scrotum derived from the superficial fascia of the abdomen. At the central tendinous point of the perinæum, it passes backwards to join the anterior layer of the triangular ligament. In cases of rupture of the urethra, this fascia prevents the urine from passing outwards upon the groin and thigh; while it allows it to mount upwards, in the loose cellular tissue of the scrotum, which it often entirely destroys.

TRIANGULAR LIGAMENT OF THE URETHRA.

The triangular ligament between the rami of the pubes is an interosseous ligament, like the membrane filling up the obturator foramen; it is connected on either side to the rami of the ischium and pubes, its base looking towards the rectum, its apex towards the sub-pubic ligament; it is pierced by the membranous portion of the urethra, which passes through the ligament *about three quarters of an inch* below the pubes. It consists of two layers, between which are situated the artery of the bulb and Cowper's glands; one layer (*the anterior*) is expanded on the bulb, keeping that body in its situation; the other (*the posterior*) is continued along the membranous portion of the urethra to the prostate gland, forms its capsule, and becomes continuous on the bladder with the vesical layer of the fascia iliaca. The ligament is sometimes called the deep perineal fascia. Urine, when it escapes from the urethra, lies under the superficial fascia, and makes its way into the scrotum. It cannot make its way into the thigh, on account of the attachment of the superficial fascia to the rami of the ischia and pubes. The triangular ligament extends for a very little distance below the urethra. In the female it is smaller than in the male.

FASCIA OF UPPER EXTREMITY

consists of tendinous fibres, which are stronger in some situations than others; it invests the entire arm, and sends partitions between the several muscles. It takes its origin superiorly from the spine of the scapula, adheres to the condyles of the humerus, and to the ridges which lead to them; passes from thence on the forearm, where it is very strong, particularly at its posterior part, and, binding down the several muscles, reaches the wrist-joint, to the annular ligaments of which it is connected.

The *palmar fascia*, of triangular form, is very strong, and takes its origin from the anterior annular ligament; from this it expands over the palm, and near the fingers divides into four fasciculi, each of which is forked and inserted into either side of the sheaths of the flexor tendons, and into the ligaments of the first phalanges.

FASCIA LATA.

The fascia lata takes its origin from the crest of the ilium, the spines of the sacrum, the os coccygis Poupart's ligament, the tuberosity of the ischium, and the rami of the ischium and pubes. From this extensive connexion it extends down the thigh, confining the different muscles in their situation and also sending partitions between them. At the posterior part of the thigh it adheres intimately to the linea aspera, and at the knee-joint to the condyles of the femur; it is then continued over the heads of the tibia and fibula, to which it adheres and forms the fascia of the leg.

Upon the anterior and upper part of the thigh, the fascia lata, from its special arrangement, has been divided into the iliac and pubic portions, and about an inch and a half below Poupart's ligament, and between the iliac and pubic portions, it presents the opening for the saphena vein. This opening is semilunar, the concavity being directed towards Poupart's ligament; it presents an internal and external cornu, and its edge, turning inwards on itself, becomes continuous with the sheath of the femoral vessels.

The *pubic portion* of the fascia lata covers the pectineus muscle, adheres to the spine of the pubes and the lineo ileo-pectinea, passes behind the sheath of the femoral vessels, and becomes *continuous* with the *fascia iliaca*.

The *iliac portion* of the fascia lata covers the sartorius, tensor vaginæ femoris, rectus and iliac muscles, and presents, towards the pubic portion, a *crescentic or falciform edge*, the aspect of which is directed dowards and inwards; the inferior cornu of this edge is continous with the outer cornu of the saphenic opening, and its superior cornu extends along Poupart's ligaments, crosses the femoral vessels, and is inserted into Gimbernat's ligament, and the linea ilio-pectinea; the upper part of the falsiform edge is called Hey's ligament.

The *cribriform fascia*. The superficial fascia, in passing over Poupart's ligament to the groin, adheres to the crescentic edge of the fascia lata, and to the edge of the saphenous opening, and is attached to that layer of the fascia transversalis which passes anterior to the sheet of the femoral vessels; this portion of the superficial fascia is perforated by numerous small bloodvessels, and by the anterior superficial absorbents of the limb, which gives it, when dissected, a cribriform appearance, from which it derives its name.

The *fascia of the leg* adheres to the heads of the tibia and fibula, and to the spine of the tibia, to the annular ligaments of the ankle-joint, and to the malleoli; it binds down the muscles, sends partitions between them, which pass from its posterior surface to the bones of the leg and interosseous membrane, and from the anterior annular ligament it is continued thin upon the dorsum of the foot.

The *plantar fascia* is very strong, and arises from the under aspect of the os calcis, is attached to the sides of tarsus and metatarsus, and sends two processes between the muscles of the sole of the foot, dividing them into an internal, a middle, and an external set. At the base of the toes it divides into five portions, each of which bifurcates, and is inserted by two fasciculi into the lateral ligaments of the joints, and into the sheaths of the flexor tendons. This fascia is strengthened by transverse fibres.

THE LARYNX.

Besides the muscles, vessels, nerves, and mucous membrane which enter into the formation of the larynx, there are four cartilages and one fibro-cartilage.

The *thyroid cartilage*, the largest, presents anteriorly a prominent angle called *pomum Adami*, which is formed by the meeting of its alæ. Each *ala* is of quadrilateral form and presents posteriorly two cornua; the superior cornu is the longest, and is connected to the great cornu of the os-hyoides by the thyro-hyoid ligament; the lesser, or inferior cornu, being connected to the side of the cricoid cartilage by synovial membrane and ligaments.

The upper margin of each ala is connected to the os hyoides by the thyro-hyoid membrane, the inferior margin being connected to the cricoid cartilage by the crico-thyroid membrane, which is of yellow colour and elastic; the outer surface of each is rough, and divided unequally by an oblique ridge, the inner surface being smooth and covered by mucous membrane.

The *cricoid cartilage* is next in size, and forms

a ring; it is narrow before and deep behind; its inferior edge is connected to the first ring of the trachea; its superior edge, anteriorly, is connected by the crico-thyroid ligament to the thyroid cartilage; posteriorly it supports the arytenoid cartilages; its inner surface is covered by mucous membrane, and its outer surface is rough, and presents posteriorly a vertical ridge for the attachment of muscles. The operation of Laryngotomy is performed in the space between the thyroid and cricoid cartilages. The crico-thyroid membrane, which is formed of yellow elastic tissue, must be incised *transversely*.

The *arytenoid cartilages*, two in number, and of triangular shape, are the smallest: the apex of each is surmounted by a small moveable cartilaginous appendix; the base, concave, moves upon the cricoid cartilage; the posterior surface, concave, lodges the arytenoid muscles, the external edge is convex for the attachment of muscles, and the inner edge is flat. The apex of each is connected to the epiglottis by a fold of mucous membrane called the *aryteno-epiglottidean fold*, and the base is connected to the cricoid cartilage by synovial membrane and ligament.

The epiglottis, resembling in form an artichoke leaf, or rather the lateral half of the kernel of the butternut, is connected by a stalk-like process to the angle of the thyroid cartilage; anteriorly it is attached to the body of the os hyoides by cellular tissue and mucous membrane, and to the base of the tongue by three folds of mucous membrane, the central one of which is called *frœnum epiglottidis;* posteriorly extend the aryteno-epiglottidean folds of mucous membrane. The dangerous disease, œdema of the larynx, is situated in the loose cellular tissue of these folds and of that of the surrounding parts. It is here that scarification may be employed according to the plan of Dr Gurdon Buck of New York.

The *glottis* is the superior opening of the larynx, and is of triangular form, its base being anterior, formed by the epiglottis, its apex, posterior and inferior, formed by the appendices of the arytenoid cartilages, and its side formed by the aryteno-epiglottidean folds.

The *rima glottidis* is also of triangular form, and placed beneath the glottis: the base is posterior, is formed by the bases of the arytenoid cartilages; the apex is anterior, corresponding to the angle formed by the alæ of the thyroid cartilage; and the sides are formed by the *inferior* or true chordæ vocales.

The *chordæ vocales*, two on either side, arise from the anterior aspect of the arytenoid cartilages, and approaching each other are inserted into the angle formed by the alæ of the thyroid cartilage: the superior is semilunar, the inferior horizontal, and between the vocal chords of either side is a small oval fossa, called the *ventricle* of the larynx; from the ventricle a pouch extends upwards between the thyroid cartilage and the superior vocal chord; it is called the sacculus laryngis.

THE THYROID BODY,

of a reddish-brown colour, consists of two lateral lobes and a connecting middle lobe. The lateral lobes are placed by the side of the trachea and larynx, and the middle lobe rests upon the anterior aspect of the two or three first rings of the trachea. Each lateral lobe is of pyriform shape, the base inferior, and the apex ascending to the thyroid cartilage; both lateral lobes overlap the carotid vessels, the inferior thyroid artery, and the recurrent nerve, and are covered by the sterno-hyoid, sterno-thyroid, and omo-hyoid muscles, the cervical fascia, and the integuments. This body or gland is supplied with blood by the superior thyroid arteries from the external carotid, the inferior thyroid arteries from the thyroid axis, which is a branch of the subclavian artery, and sometimes by an artery from the arteria innominata, or from the aorta itself, called the middle thyroid artery; its blood is returned by the thyroid veins, which descending on the anterior aspect of the trachea empty themselves into the left vena innominata. No excretory duct has been discovered emerging from this body. The names of Goitre and Bronchocele have usually been given to enlargement of the thyroid gland.

HERNIA.

Before commencing the study of Inguinal, and more especially that of Femoral Hernia, it is absolutely necessary that the student should be well acquainted with the anatomy of the os innominatum, and particularly with that of the pubic and iliac portions of that bone. It is also of great advantage to obtain a pelvis, upon which Poupart's ligament is well preserved, with that portion of it to which the name of Gimbernat has been applied. The distance from the anterior superior spinous process of the ilium to the symphysis pubis, in most subjects, is about six inches. *The Spine* of the pubes is an inch and a quarter from the symphysis. From this spine, a sharp ridge or border extends obliquely backwards and outwards, and is called the linea ilio-pectinea, and which in the subject is covered by a ligamentous expansion—an inch and a quarter, on the outside of the spine of the pubes is a depression on the upper face of the bone, upon which we find the femoral absorbents, the femoral vein, and outside of the vein, the femoral artery.

Poupart's ligament extends from the anterior superior spinous process to the spine of the pubes, where it is inserted; a portion of this ligament extends backwards and inwards, and is inserted into the linea ileo-pectinea. This is called Gimbernat's ligament. It presents externally a sharp lunated border, which, looks towards the femoral vein.

HERNIA.

Hernia is a protrusion of an organ from the cavity in which it is naturally placed All the great cavities of the body, as that of the cranium, thorax, abdomen, and pelvis, are lined by a serous membrane, which is protruded before the organ, as it is escaping from its cavity, and is called the hernial sac. Hernia may therefore occur from any of the great cavities; but for several reasons is most commonly met with in the region of the abdomen. To understand the nature of abdominal hernia, it is necessary to consider the structure of the walls of the abdomen, and the organs contained in that cavity. In looking on the inside of the anterior wall of the abdomen, we find it smooth and polished, which appearance is due to the serous lining, or peritoneum. On stripping off this membrane we expose to view a strong, dense white fascia, the fascia transversalis, so called because it lines the posterior surface of the transversalis muscle. This fascia is of very great extent; commencing below, we find it attached to the whole extent of Poupart's ligament, to the internal lip of the crest of the ilium; it becomes weak and thinner towards the linea alba; superiorly it may be traced as high as the diaphragm. From the great extent of surface of this membrane, lining as it does the whole anterior and muscular wall of the abdomen, and having the peritoneum attached to its posterior surface, it is evident that any organ or viscus which tends to push the peritoneum before it, *must* also protrude the fascia transversalis. This fascia therefore must necessarily form a covering to all kinds of abdominal hernia. The next layer, passing from within outwards, is the muscular wall of the abdomen, consisting of five pair of muscles with their tendons, their fibres running in various directions, some downwards, some upwards, and others transversely, while some again are arranged vertically. By the decussation of these fibres in so many various directions, the muscular and tendinous wall of the abdomen is rendered very strong and resisting; and were it as much so in all points as has been described, it is difficult to conceive how a hernia could ever take place. Upon the outer surface of the muscles we find the superficial fascia and fat, and upon this the skin. The *posterior* wall of the abdomen is formed by the lumbar vertebræ, psoas magnus quadratus lumborum, and superiorly the abdominal cavity is bounded by the diaphragm, inferiorly it communicates with that of the pelvis. The viscera of the abdomen, which are most likely to protrude and form hernial tumors, are those which are endowed with the greatest mobility. Accordingly we find that the small and large intestines, and the omentum, are most frequently found in hernial protrusions. In a few rare cases nearly all the viscera have been seen in a hernial sac, with the exception of the kidneys, pancreas, and duodenum. From the description of the great strength of the muscular, tendinous, and fascial structures, entering into the formation of the abdominal wall, it might be asked, how is it ever possible for hernia to occur? We shall find that it does so only at certain points, which are congenitally defective, or which are weakened by the transit of certain organs, which in their progress of development are necessarily obliged to pass through the anterior abdominal wall. It now becomes necessary to examine the situation and nature of these various openings, from which the different varieties of hernia protrude.

If a hernial tumour presents at the superior part of the groin, above Poupart's ligament, passing down in the scrotum and towards the testicle, it is called Inguinal Hernia. If the tumour shall have followed the course of the spermatic cord, it is called oblique; if it has not done so, but passed directly out of the cavity of the abdomen, passing out of the external ring, it is then called Direct Inguinal Hernia. *The anatomy of Hernia can never be comprehended, without a careful dissection of the dead body.* The mode of making the dissection will therefore here be described, and *in the simplest manner*, for the benefit of those who are just entering upon their anatomical studies. A block being placed under the loins of the subject, and the thighs widely separated, the legs lying over each side of the table, make an incision from the umbilicus to the symphysis pubis, another from the umbilicus outwards towards the lumbar vertebræ. This incision should be made carefully through the skin and superficial fascia, down to the tendon of the external oblique. The superficial fascia should be cautiously raised from the tendon, and the flap dissected back towards the crest of the ilium; when some of the muscular fibres of the external oblique will be displayed. On approaching Poupart's ligament, the superficial fascia will be found closely adhering to it, and can be separated or detached with the handle of the knife; the superficial fascia is formed of condensed cellular tissue and fat, and has several small vessels ramifying in its substance. The tendon of the external oblique is now fairly exposed, the direction of its fibres is downwards; at the upper part of the groin, the tendon becomes thickened, and forms a strong tendinous band, called Poupart's ligament, which extends from the *anterior superior spinous process* of the ilium, to the *spine* of the *pubes*. A rounded reddish cord is seen emerging from the abdomen through an opening in the tendon of the external oblique. This cord is the spermatic cord, which consists of the vas deferens, which is the excretory duct of the testicle; the spermatic artery and vein, the cremasteric artery, the artery of the vas deferens, the nerves of the testicle, the absorbents, and the cremaster muscle, which covers or is spread out over the other constituents of the cord. The opening in the external oblique tendon, through which the cord passes, is the *external abdominal ring*. A thin fascia proceeds from the edges of the ring, downwards upon the cord, and is called intercolumnar fascia. When this fascia is dissected away, the ring and cremaster become distinctly seen. The "ring" is now seen not to be circular, but triangular.

The portion of tendon which passes above the cord is called the superior column of the ring; it passes over the symphysis pubis, and interlaces with its fellow of the opposite side. The portion of tendon passing below the cord is inserted with Poupart's ligament into the spine of the pubes, and is called the inferior column of the ring. It is often important, as in the examination of recruits for the army or navy, &c., to ascertain whether a tendency to a hernia exists, and therefore to know the exact place at which such tumour will protrude, or in [other words the situation of the external ring. This is about one inch or an inch and a quarter from the centre of the symphysis pubis, and may be easily found, by grasping the cord with the fingers and thumb, and tracing it up to the ring. It now becomes necessary to trace the course of the cord, from the ring upwards, and to ascertain the relation of the surrounding parts to it. An incision is to be made through the external oblique tendon, commencing near the outer extremity or attachment of Poupart's ligament, and carried parallel with the ligament (and about a quarter of an inch above it) almost to the external ring, but so as to preserve the ring, and then turning suddenly upwards towards the median line.

The flap is then to be raised, when the muscular fibres of the internal oblique are brought into view, as well as the fibres of the cremaster, which originate from it, and also from Poupart's ligament. The muscular fibres of the internal oblique and transversalis arise from the outer half of Poupart's ligament, arch over the cord, and their fibres become united, forming the conjoined tendon, which passes inwards and downwards, and is inserted into the linea ileo-pectinea, and into Gimbernat's ligament, hereafter to be described. If the internal oblique and transversalis are now carefully detached from Poupart's ligament, the fascia transversalis is seen as a dense whitish membrane, the muscles may be still further raised up, when the fascia can be traced to the outer edge of the cord, and from thence to the edge of the rectus muscle, gradually becoming very thin and delicate towards the median line. On first inspection the cord appears to pass through an opening in the fascia transversalis; this point is called *the internal abdominal ring*, and is just half way between the anterior superior spinous process of the ilium and the symphysis pubis, and half an inch above Poupart's ligament. The ring, however, is not an *opening*, but a funnel-shaped process of the fascia, sent down upon the cord; and it is in this funnel that the hernia engages as it descends.

COVERINGS OF OBLIQUE INGUINAL HERNIA.

As the intestine or omentum protrudes from the abdomen, it pushes before it: 1. The peritoneum. 2. It enters the funnel-like process of the fascia transversalis, and is covered by it. 3. It passes under the fibres of the cremaster, and is covered by that muscle. 4. It is next invested by the intercolumnar fascia. 5. Its last or external covering is from the superficial fascia and skin.

OF THE EPIGASTRIC ARTERY.

This artery is a branch of the external iliac, and is given off from it close to the upper margin of Poupart's ligament, and sometimes behind it. At first it passes slightly downwards, then upwards and inwards, towards the edge of the rectus muscle, to which it sends many branches. It inosculates with the internal mammary. It lies on the *inner* edge of the internal abdominal ring, and behind the fascia transversalis; consequently when the intestine descends in oblique hernia, following necessarily the course of the cord, *it must push the epigastric artery to the inside.*

OF THE INGUINAL CANAL.

This is merely the space which the cord occupies between the external and internal abdominal rings. Anteriorly, it is bounded by the tendon of the external oblique, posteriorly by the conjoined tendon and fascia transversalis, and inferiorly by the upper grooved border of Poupart's ligament. The obliquity of this canal is a considerable protection against the occurrence of hernia, which would doubtless have been much more frequent had the cord passed *directly* from the cavity of the abdomen. When the abdominal muscles are in strong action they act as a valve pressing together the sides of the inguinal canal, and thus tend to prevent the descent of the viscera.

DIRECT INGUINAL HERNIA.

The conjoined tendon formed by the internal oblique and transversalis muscles, passes behind the outer edge of the rectus muscle, to be inserted into the linea ileo-pectinea, and into Gimbernat's ligament. This conjoined tendon is closely attached to the inner portion of the fascia transversalis, or, in other words, to that part of it which extends from the inner margin of the internal ring, inwards towards the rectus muscle, and downwards towards the pubes. This close attachment of the conjoined tendon to the inner portion of the transversalis is of great importance, as preventing the occurrence of direct hernia, inasmuch as the powerful contraction of the internal oblique and transversalis upon the conjoined tendon tightens and braces the fascia transversalis, to which it is attached. On examination it will be found that the conjoined tendon and fascia transversalis close up and protect a triangular space, between the epigastric artery, the outer edge of the rectus, and the pubes below. In most subjects this combination of fascia and tendon is strong enough to resist the tendency of the viscera to protrude when violently compressed by the abdominal muscles; but in some subjects these parts are naturally imperfect, or so weak as to be incapable of resistance to the passage of a hernial tumour. When this protrudes through this deficiency in the conjoined tendon and fascia it affords an example of direct inguinal hernia. As we trace the progress of the hernia, we find that it continues to descend to the external ring, through which it passes, receiving as it goes through that ring the intercolumnar fascia, and then the superficial fascia and skin. It will thus be seen, that this variety of hernia is well named, inasmuch as its course is straight forwards or *direct*, from the deficiency in the conjoined tendon to the external ring. It differs from oblique inguinal hernia, as it leaves the cord to the outside, and is not covered by the cremaster—as it passes down, it pushes or leaves the epigastric artery to the *outside*. The form of the direct hernial tumour is *generally* rounded, that of the oblique most usually pyriform.

COVERINGS OF A DIRECT INGUINAL HERNIA.

1. The peritoneum. 2. The fascia transversalis. 3. Intercolumnar fascia. 4. Superficial fascia and skin.

OF THE STRICTURE.

1. The stricture may be caused by the external abdominal ring. 2. By the constricting edges of the internal oblique and transversalis muscles, where they arch over the cord. 3. By the edges of the opening in the conjoined tendon. 4. More frequently at the internal ring, or in the neck of the sac itself. In all cases of strangulated hernia, it is safest to divide the stricture directly upwards.

TAXIS.

The patient should lie upon his back, a pillow should be placed under the pelvis and another under the shoulders, the thighs should be raised to a right angle with the body, and the knees brought close together. In applying the taxis in oblique hernia, the pressure on the tumour must be made in the direction of the course of the cord or towards the anterior superior process of the ilium. In direct hernia the pressure may be made directly upwards and backwards in accordance with the direction in which the hernia came down. In using the taxis, great gentleness and caution are to be observed.

FEMORAL HERNIA.

Supposing that the dissection for inguinal hernia has been made on the left side, continue the incision from the spine of the pubes downwards for five inches in a perpendicular direction, from the termination of which a second incision is to be made across the fore part of the thigh so as to allow a flap of the superficial fascia and skin to be reflected outwards. In raising up the superficial fascia, the facial lata, on which it reposes, will be exposed. This must be done *with caution*, so as to avoid injuring the saphena vein, or the fascia lata—frequently using the handle of the knife to separate the parts. A number of lymphatic glands or vessels, and small arteries and veins, are involved in the layers of the superficial fascia. Commence the separation of the superficial fascia from the fascia lata; from the inner portion of Poupart's ligament and the pubic part of the fascia lata; lower down, find the saphena vein, separate the superficial fascia from the vein, which is to be left, reposing upon the fascia lata. This vein passes up from the inner part of the foot and leg, to join the femoral vein at the upper and inner part of the thigh.

By now gently pushing with the handle of the knife on the pubic side of the fascia lata, from above downwards and inwards, upwards under the vein, and upwards and to the outside of it, we bring into view a *lunated* edge of an opening in the fascia lata. This edge or border is to be carefully traced upwards and outwards; when, if we continue cautiously to separate as before, it will have the appearance of a *falciform* margin or *process*, passing up to be inserted into the inner portion of Poupart's ligament. The lunated and falciform edges or borders above described, will be found to form the margin of an opening in the fascia lata for the passage of the saphena vein, and hence called *the saphenous opening.*

The femoral artery and vein, side by side, lie in a sheath of areolar or cellular tissue of considerable thickness, which occupies the saphenous opening. The femoral vein is on the *inside* of the artery. This sheath is closely connected with the superficial fascia of the groin which covers over the saphenous opening, adhering closely to its margin. This superficial fascia is now to be removed from the saphenous opening, but in such a manner as to leave a large portion of cellular and adipose tissue, covering the vessels, and which in fact constitutes the *anterior* part of the sheath of the vessels above mentioned.

The fascia lata is now seen as a dense, strong, fibrous membrane, covering the muscles on the upper portion of the thigh. Its outer part, or *iliac portion*, is attached to the crest of the ileum, and to the whole inferior border of Poupart's ligament, and is on a higher plane than the pubic portion, which is attached to the symphysis, spine of the pubes, and linea ileo-pectinea. By extending the thigh and throwing it strongly outwards, Poupart's ligament and the upper portion of the fascia lata are put strongly upon the stretch. On the other hand, by flexing the thighs towards the abdomen and turning the knee inwards, the same parts become greatly relaxed. This fact is important to be remembered in the reduction of hernia.

The abdomen may now be opened by cutting in the line of the first median and transverse incisions into that cavity. The peritoneum may now be cautiously separated from the fascia transversalis, beginning on the outside, near the middle of the crest of the ilium, and passing inwards towards the median line. The fascia transversalis is attached to the crest of the ilium, to the whole extent of Poupart's ligament, *and can be traced under Poupart's ligament to become continuous with the cellular and adipose tissue which covers the femoral artery and vein*, as they lie in the saphenous opening. The peritonæum is now to be detached from the parts in the iliac fossa, when a strong pearly white fascia is exposed which is called the *fascia iliaca*. The fascia iliaca covers over the psoas magnus and iliacus internus muscles; it is attached to the inner lip of the crest of the ilium where it meets the transversalis fascia *in a seam;* also in a similar manner it is attached to the outer half of Poupart's ligament; it then passes *underneath* the external iliac artery and vein, *these vessels reposing upon it.* It can now be traced, under Poupart's ligament, becoming continuous with the pubic portion of the fascia lata of the thigh. The fascia transversalis dips down on the inside of Gimbernat's ligament to join the fascia iliaca. It will thus be seen that by the fascia transversalis, passing down from the abdomen over the anterior surface of the femoral vessels as they pass under Poupart's ligament, and by the passage of the fascia iliaca underneath these vessels, a large funnel-like sheath is formed, in which the vessels are enclosed, and which is called the "*sheath of the vessels*," or the "*femoral sheath*." The finger may now be inserted on the *inside* of the external iliac vein between the vein and the sharp lunated border of Gimbernat's ligament, when it will easily pass into an opening called the *femoral ring*. The finger is now in the funnel-like sheath of the vessels, and by a little pressure passes easily under Poupart's ligament, still covered by the anterior part of the sheath which was left on the vessels when the superficial fascia was removed from over the saphenous opening. It is through the femoral ring that a femoral hernia descends from the abdomen, passing down into the femoral sheath, pushing the *anterior* layer of the sheath (*which is the fascia transversalis*) before it, and then emerging from the saphenous opening upon the anterior and upper portion of the thigh, where it is then covered by the superficial fascia and skin. While the finger is in the ring, it will be found to pass most readily downwards and forwards, which is the general course of a femoral hernia, sometimes turning upwards over Poupart's ligament; while the finger of an assistant is passed into the ring and femoral sheath, this last may be opened, to understand more readily how the sheath envelops or encloses the hernia. The fascia iliaca may now be raised up from the iliacus internus and psoas magnus muscles, and cut away from its attachment to the crest of the ilium and Poupart's ligament. The fossa of the iliac bone will be seen to be filled up by the above named muscles; next we find passing towards the median line, the external iliac artery, next the external iliac vein, then the femoral ring; after that we come to the lunated border of Gimbernat's ligament. We thus perceive that all the space between the anterior superior spinous process of the ilium and the femoral ring, and between the iliac fossa and Poupart's ligament, is occupied by muscles and by the femoral vessels, and strengthened by the meeting together of the fascia transversalis and fascia iliaca (in a seam) from the crest of the ilium to the outer side of the external iliac artery. As the fascia transversalis passes down upon the vessels, and while under Poupart's ligament, it sends down a partition or septum of fascia, between the external iliac artery and vein, and another, which passes down on the inside of the vein, forming the outer wall of the femoral ring. The inner wall of this ring or opening is formed by a septum or partition of fascia, passing downwards from the fascia transversalis to the fascia iliaca, and lining the lunate border of Gimbernat's ligament. It thus becomes evident that the parts which separate and shut the abdomen from the thigh, are so arranged that there is only one place where femoral hernia is likely to occur, and that is at the point already designated as *the femoral ring*. In the natural condition of the parts this ring is occupied by a small absorbent gland, lymphatics, and a small quantity of cellular and adipose tissue.

OF THE EPIGASTRIC ARTERY.

This artery ordinarily arises from the external iliac close to the upper margin of or behind Poupart's ligament. Its course is at first slightly downwards, then upwards and inwards, to the outer border of the rectus muscle. When a femoral hernia descends, the epigastric is pushed to the *outer* side of the hernial tumour. Occasionly the obdurator, instead of coming off, as it usually does, from the internal iliac, arises by a common trunk with the epigastric from the external iliac; and in such cases the hernia is almost encircled by the artery. This variety is said by Cloquet and others to occur once in about three or four cases; but in the majority of instances the tumour, in its descent, would probably evade the artery, which would be pushed to the outside and below the hernia. The possibility, however, of the hernia becoming encircled by the artery should always be kept in mind, and should suggest great caution in dividing the stricture, cutting or rather *nicking* it only so much as is just sufficient to liberate the constricted part.

OF THE STRICTURE.

A femoral hernia may be constricted. 1. In the femoral sheath and by the falciform edge of the fascia lata, as it passes up to be inserted into Gimbernat's ligament. 2. By the posterior edge of Poupart's ligament. 3. More frequently in the neck of the sac itself. As the hernia passes through the femoral ring it is greatly constricted, this ring being bounded on the inside by the sharp lunated border of Gimbernat's ligament, superiorly by the posterior edge of Poupart's ligament, inferiorly by the *fibrous covering of the os pubis* and pubic portion of the fascia lata, while the *outer* boundary is the femoral vein: and it is only in this direction that the ring admits of much dilatation. As the hernial tumour is nearly surrounded by dense and fibrous tissues, delay in relieving the strangulation is more dangerous than in inguinal hernia.

COVERINGS OF A FEMORAL HERNIA.

As it descends, entering the femoral ring, it pushes before it—1. The peritoneum. 2. A small lymphatic gland, absorbent vessels, cellular tissue, and fat, which fill up the femoral ring, and together constitute what has been called the *septum crurale*. 3. The transversalis fascia, or the anterior wall of the funnel-like or femoral sheath. 4. The superficial fascia and skin.

TAXIS.

In applying the taxis the abdominal muscles

must be relaxed as much as possible, the thighs are flexed towards the abdomen, the knees turned inwards. The hernia should be pressed gently backwards, as it were into the thigh, and then upwards. If it has turned up over Poupart's ligament, it must be carefully disengaged from it, brought down, and then pressed backwards and upwards as before. Femoral hernia is usually very small, especially when recent. It is often mistaken for an enlarged lymphatic gland, and such errors are not unfrequently fatal. If a patient, and especially a female, is seized with nausea and vomiting, pain in the abdomen, and obstinate constipation, the possibility of the existence of a strangulated hernia should never be forgotten, and the necessity of an immediate and careful examination of the points where hernia (and particularly femoral) may occur, should be insisted upon. From false motives of delicacy, but more frequently from ignorance, an examination has not been made; or if so, the tumour, which is generally small, if recent, has been mistaken for a lymphatic gland, and poultices have sometimes been applied to what was supposed to be a suppurating tumour, which being opened by an incision, fæces were discharged from the wound, the patient dying soon afterwards. Another very dangerous error is that of mistaking crural for inguinal hernia, inasmuch as the mode of applying the taxis and of operating, is essentially different in the two kinds of hernia.

Diagnosis.—The neck of the hernial tumour is situated above Poupart's ligament in inguinal hernia. In femoral hernia it is below, and if the tumour is drawn down in femoral hernia, Poupart's ligament may be traced above it. The neck of the tumour in inguinal hernia is above the spine of the pubes, that of femoral hernia is below and to its outside. Femoral hernia may also be mistaken for psoas abscess, and for other diseases, but the history of the case and a careful examination of the parts will generally suffice to determine the character of the disease.

The student may now remove the cellular and adipose tissue from the femoral artery and vein, in such manner as to get a clearer and more distinct view of the saphenous opening. He may then remove all the cellular and adipose tissue from the vessels, above and under Poupart's ligament, when he will obtain a more perfect idea of Gimbernat's ligament, and of the manner in which the fascia iliaca passes under the great vessels to become continuous with the pubic portion of the fascia lata.

The parts in hernia should be carefully and repeatedly dissected, and may be compared with the description here given. The knowledge obtained from reading alone is worse than useless, and it would be much better for the student never to read a word upon hernia until, with scalpel in hand, and his book open before him, he commences his dissection upon the subject.

www.ingramcontent.com/pod-product-compliance
Lightning Source LLC
Chambersburg PA
CBHW031934230426
43672CB00010B/1925